A Practical Companion to Spanish

With Nautical and Aeronautical Science Vocabulary and Practice Activities

Robert L. Rankin, MBA

Preface:

The Spanish language is rich in many facets, and for those who have a desire to understand what native Spanish-speakers and other speakers of Spanish are saying is a worthy cause. The world opens up to us, as we open up to others, and the most beautiful way to grow is in language and our communications amongst one another. The feelings that we feel become much more intense as we begin to understand the feelings of others, and we are all the better for doing so.

This is for the student of any age whom has interactions, whether frequent or sparse, with Spanish speaking people. It is our sincerest hope that this becomes a manual to practically immersing yourself in the language, as it is essential to do so in order to see, hear, write, and eventually, when you become comfortable enough, to speak.

The design of this manual is for **simplicity.** There are more advanced topics in the Spanish language that are not covered in this manual, and if you are ever interested in going beyond this manual, **we recommend asking a native Spanish speaker to tutor you.**

The following pages are the efforts of many more than one native Spanish speaker over many wonderful friendships over many cherished years. It is our hope that you too will share in the wealth of knowledge that understanding Spanish can bring.

Introduction:

This is an approximately 8 week program designed to introduce students and professionals to the Spanish language, particularly those whom are employed in the nautical and aeronautical sciences. This was written in the style of a manual that one would find in any of the various products one purchases from overseas or nationally. Our sincerest hope is that you find friends along your journey into the Spanish language and learn alongside them to enrich your life and live la vida sana.

The manual is taken into lessons, each of which may be referenced to after the completion of each lesson for ready-to-use communications which are frequently used in the language. The lessons concerned with the conjugations of Spanish action words are emphasized and are the fundamental building components to communicating Spanish. For this reason, **the first two lessons in this manual are conjugation tables** which must be mastered preceding further lessons.

Lesson 3 deals with the pronunciation of the Spanish language, since we anticipate our readers desire to speak the language in various circumstances. Lesson 3 also presents the reader with a quick-reference to phonetic differences in the English and Spanish alphabets. This tool is written for those who have an elementary understanding of Spanish and only wish to improve their speech in Spanish as well as those who have an elementary understanding of English and only wish to improve their speech in English. Our objective is to provide a cross-reference to the reader for personal and professional use.

Lesson 4 considers the numeration and nomenclature of the Spanish numbering system with the pronunciations of each number and their successive pronunciations into the teens, bases of ten, hundreds, and thousands. The discussion of time in the Spanish language is also presented as a how-to with the pronunciations phonetically of days of the week, months of the year, and general times of the day like morning, mid-day, evening, and night.

Lesson 5 informs the reader of the imperfect form, how to conjugate action words in the imperfect form to say what was happening or what something or someone was doing in the past. Lesson 5 also provides the reader with the tools to form gerunds in Spanish, or words ending with -ing to describe a particular action in the past, present, or future. Lesson 5 also presents the past-participle, or any words ending with -ed to describe a particular action that has already been completed. The combination of the imperfect conjugation, gerunds, and past participle provide the reader an ability to phrase past and present actions.

Lesson 6 is all about completed actions and how to say what had already been done. When reading Spanish, one will frequently encounter the conjugation of Haber, meaning "to have", as well as the past participle to inform the reader that an action has been completed or had been completed definitively. The reader will find how to conjugate the six forms of Haber in order to describe what any subject had done in the past, recalling how to form the past participle in Lesson 5.

Lesson 7 is future-directed and provides the toolkit for the conjugation of future-tense action words. The future tense is used to inform persons of your intentions in the future or of the subject's intentions in the future. The conjugation of the irregular action word "Ir" is also provided here with the formula to provide clarity between general future phrasing and definitive future phrasing. This lesson provides the difference between "I will ___." And "I am going to ____."

Lesson 8 is the most useful in this manual, as it is a subjective reference to the Spanish vocabulary and encourages the reader **to read**, **pronounce**, and **write in past, present, and future tenses** learned in each of the previous lessons with an independent skills assessment to be taken at the readers' leisure to re-solidify the weekly lessons.

In addition, a list of the largest nautical shipbuilding companies and the largest aerospace entities have been listed as a reference to aid the reader. Such entities are of importance when considering the international trade climate, military operations, and general international travel. We encourage the reader to look into these entities to further their knowledge of the nautical and aeronautical sciences.

Because nautical and aeronautical science is practiced all over the earth, we hope this reference provides a deeper understanding of how far-reaching the nautical and aeronautical sciences are and what their theories and laws can teach us about the natural world in which we live.

Let us begin, shall we?

Table of Contents

Preface 3

Introduction 4

Lección 1 : CONJUGATIONS OF REGULAR VERBS 8

Identifying infinitives and overview of the 6 Conjugation forms of Spanish 8

Differentiating the 6 Conjugation forms of Spanish + Practice 9

Matching game activity for the 6 Conjugation forms of Spanish 10

Activity answers 11

Lección 2: CONJUGATIONS PRACTICE AND PRONOUNS 12

The Present Tense Conjugation of -AR Verbs + Practice 12

The Present Tense Conjugation of -ER Verbs + Practice 13

The Present Tense Conjugation of -IR Verbs + Practice 14

Lección 3: PRONUNCACION 17

Don't Stress It 17

Abecedario/Alphabet + Examples in English 18

Trilling+Rolling an R Method 20

Lección 4: QUE HORA ES? What time is it? 21

Spanish Numerals and How to Count to the Trillions 21

Telling the Time of Day in Spanish 23

Telling the Day of the Week in Spanish 23

Morning, Noon, Night 24

Lección 5: ESTABAMOS PRACTICANDO. We were practicing.   **25**

The Imperfect Tense Conjugation of -AR Verbs + Practice   **25**

The Imperfect Tense Conjugation of -ER Verbs + Practice   **26**

The Imperfect Tense Conjugation of -IR Verbs + Practice   **27**

Forming Gerunds in Spanish (endings with "ing") + Practice   **28**

Forming the Past Participle in Spanish (endings with "-ed") + Practice   **29**

Recap   **30**

Lección 6: PRETERITO PERFECTO and completed actions   **31**

Forming the Past Perfect Tense of Spanish Verbs (Formula)   **31**

Example for forming Past Perfect Tense for -AR Verbs   **32**

Example for forming the Past Perfect Tense for -ER Verbs   **32**

Example for forming the Past Perfect Tense for -IR Verbs   **33**

Practice   **33**

Lección 7: FUTURO for future actions   **34**

Forming the Future Tense of Spanish -AR Verbs with Addition + Practice   **34**

Forming the Future Tense of Spanish -ER Verbs with Addition   **35**

Forming the Future Tense of Spanish -IR Verbs with Addition   **35**

The Informal Future Tense of Spanish + Examples   **36**

Lección 8: PRACTICAS VOCABULARIO + Recap   **37**

YOU HAVE THE TOOLS AND KNOW THE RULES, GO GET EM TIGER!   **37**

-VOCABULARIO REGULAR Y SOCIAL/ REGULAR AND SOCIAL VOCABULARY   **38**

-VOCABULARIO DE NAUTICAL / NAUTICAL VOCABULARY   **40**

-VOCABULARIO AERONAUTICA / AERONAUTICAL VOCABULARY   **74**

LAS PRINCIPALES ENTIDADES DE CONSTRUCCIÓN NAVAL Y AEROESPACIAL/TOP SHIPBUILDING AND AEROSPACE ENTITIES/   **115**

LECCION 1 : CONJUGATIONS OF REGULAR VERBS

In Español, there exist (6) modification of ALL action words. Action words like, "searching", "seeing", "tasting", "smelling", "touching", and "feeling" have a raw, unmodified form known as an INFINITIVE.

ALL ACTION WORDS FIT INTO ONE OF FOUR CATEGORIES:

1. **-er** action words (Examples: Contest**er**, Aprend**er**, Met**er**)
   Any Spanish action word that ends in -er.
2. **-ar** action words (Examples: Sac**ar**, Qued**ar**, Entr**ar**)
   Any Spanish action word that ends in -ar.
3. **-ir** action words (Examples: Discut**ir**, Ven**ir**, Permit**ir**)
   Any Spanish action word that ends in -ir.
4. **Irregular action words (Dar, Querer, Ir, Tener)**
   Any Spanish action word whose conjugated endings do not follow the regular -er, -ar, ir conjugations.

WHAT IN THE WORLD IS A CONJUGATION, AND WHY IS IT IMPORTANT TO ME?

Depending on the subject you wish to describe or what you wish to communicate, the action word you use (**the infinitive**) and the ending of the modified action word (**the conjugation**), will provide the reader or listener information to whom or for whom the action word is referring.

For example, in English, it is improper to write infinitive phrases at the beginning of a sentence : **"To bring the book"**, **"To write the page"**, and **"To find the answer"** are all incomplete sentences. Properly, in English, we may write or say, **"I bring the book."**, **"You write the page."**, and **"They find the answer."**

This provides the clarification of whom or what the action words **"bring"**, **"write"**, and **"find"** are referring to. **Spanish is the same**, grammatically, because without the conjugated form, the listener or speaker does not know of whom or what the action refers to.

You will find a tool on the following page to assist you with your mastery of the Six Conjugation Forms of All Spanish Verbs! □

THE SIX CONJUGATION FORMS OF ALL SPANISH VERBS

Yo form (I _____.)	Nosotros form (We _____.)
Tu form (You _____.)	Vosotros form (You all _____.)
Usted, Ello, Ella form (You [formal], He, She ___.)	Ustedes, Ellos, Ellas form (You [plural formal], They (Male), They (Female)__.

The **Yo form** in Spanish is the equivalent to saying "**I _____.**"

The **Tu form** in Spanish is the equivalent to saying "**You _____.**" This is the familiar form which is used among friends.

The **Usted, Ello, Ella form** in Spanish is the equivalent to saying "**It, He, She _____.**" This is the formal form which is used in professional environments.

The **Nosotros form** in Spanish is the equivalent to saying "**We _____.**"

The **Vosotros form** in Spanish is the equivalent to saying "**They all ___.**" This is the informal, **familiar form** and is not commonly used in spoken or written Spanish. For this reason, we will not focus on the vosotros form in this manual.

The **Ustedes, Ellos, Ellas form** is the equivalent to saying "**They all, the males, the females _____.**" This is the formal form which is used in professional environments.

Say Aloud: Please say the names of the Spanish form words- **Yo, Tu, Usted, Ello, Ella, Nosotros, Vosotros, Ustedes, Ellos, Ellas-** ten times each as well as their equivalent in English- **I, You(familiar), It, He, She, We, They all(familiar), They all(Professional), Males, Females.**

Write Right: Please write the Six Conjugation Forms of All Spanish Verbs four times in the blanks below:

I: _____ You_____. He_____.
She_____. It(Professional)_____.
We_____.

They all (familiar)_____.
They(Male plural)_____.
They(Female plural)_____.
They all(Professional)_____.

Now please complete the matching game:

Nosotros	I
Ustedes	She
Ellas	You (familiar)
Yo	They all (familiar)
Ello	We
Ella	It
Vosotros	Males
Tu	Females
Usted	They all (Professional)
Ellos	He

Once More, please complete the matching game:

Vosotros	He
Ustedes	We
Ello	They all (Professional)
Ellas	She
Usted	I
Tu	They all (Familiar)
Nosotros	It
Ellos	You (Familiar)
Yo	Females
Ella	Males

Check your answers on the following page.

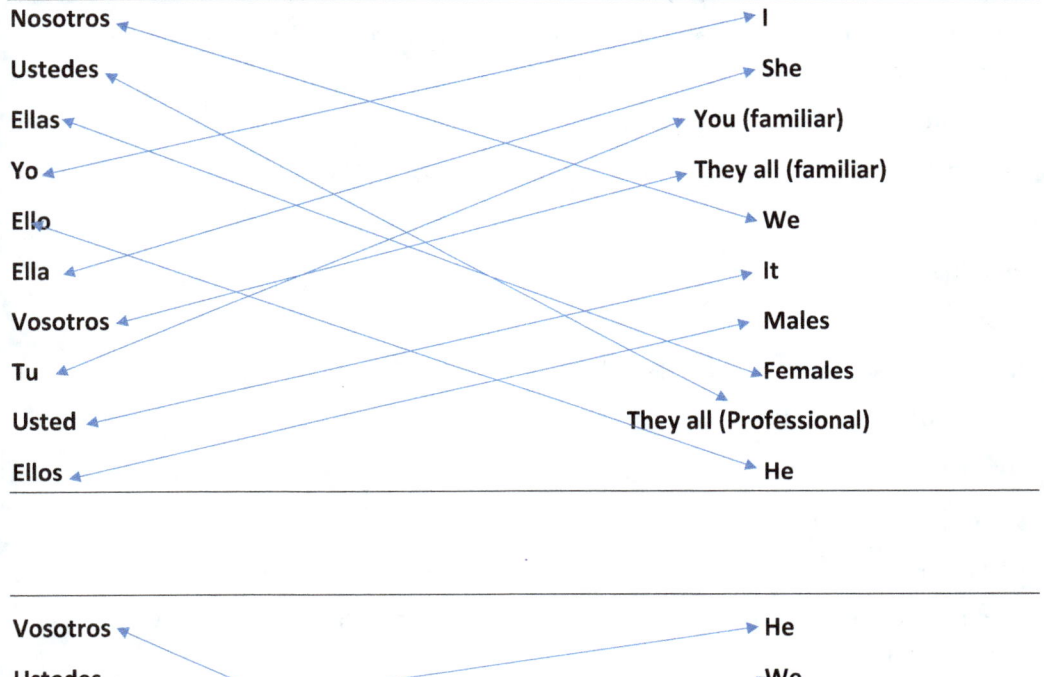

Nosotros	I
Ustedes	She
Ellas	You (familiar)
Yo	They all (familiar)
Ello	We
Ella	It
Vosotros	Males
Tu	Females
Usted	They all (Professional)
Ellos	He

Vosotros	He
Ustedes	We
Ello	They all (Professional)
Ellas	She
Usted	I
Tu	They all (Familiar)
Nosotros	It
Ellos	You (Familiar)
Yo	Females
Ella	Males

Recall that Spanish has **six conjugations**. The reason for this is that Spanish has **six pronouns** to give possession of the action which are:

Yo: "I/Me"

Tu: "You"

Usted: "Formal you/It"

Ello: "He"

Ella: "She"

Nosotros/Nosotras: "We" (Male/Female)

Vosotros/Vosotras: "All of you" (Male/Female)

Ustedes: "Formal all of you"

Ellos: "They" (Male)

Ellas: "They" (Female)

Helpful hint: The ending of the pronouns in Spanish are either "**o**" or "**a**" to clarify if the subject is **male** or **female**.

The following table is the **Present Tense** conjugation for **ALL REGULAR -ar verbs** in Spanish using the infinitive **Hablar**, meaning to speak:

Pronoun	Infinitive	Remove -ar	Add ending	Conjugation
Yo (I)	Hablar	Habl	-o	Hablo
Tu (You)	Hablar	Habl	-as	Hablas
Usted (It)	Hablar	Habl	-a	Habla
Nosotros (We)	Hablar	Habl	-amos	Hablamos
Vosotros (They all)	Hablar	Habl	-ais	Hablais
Ustedes (They)	Hablar	Habl	-an	Hablan
Ellos (Males)	Hablar	Habl	-an	Hablan
Ellas (Females)	Hablar	Habl	-an	Hablan

The recipe for **-ar verbs** is: **remove the -ar end of the infinitive** and **add the proper ending**, depending on who or what is performing the action. All conjugations, except for irregular verbs follow this pattern.

Please write the following five times each:

Hablo:_____

Hablas:_____

Habla:_____

Hablamos:_____

Hablais:_____

Hablan:_____

Please write the proper conjugations for the following phrases five times:

I speak: _____.

You speak:_____.

They,He,She speaks:_____.

We speak:_____.

They all speak:_____.

They, Males, Females speak:_____.

*You now can conjugate the **present tense of all regular -ar Spanish verbs**!!*

Find a **regular Spanish verb ending in -ar** and write the **six conjugations** of the verb:

_____

_____

_____

_____

The following table is the **Present Tense** conjugation for **ALL REGULAR -er verbs** in Spanish using the infinitive **Comer**, meaning to eat:

Pronoun	Infinitive	Remove -er	Add ending	Conjugation
Yo (I)	Comer	Com	-o	Como
Tu (You)	Comer	Com	-es	Comes
Usted (It)	Comer	Com	-e	Come
Nosotros (We)	Comer	Com	-emos	Comemos
Vosotros (They all)	Comer	Com	-eis	Comeis
Ustedes (They)	Comer	Com	-en	Comen
Ellos (Males)	Comer	Com	-en	Comen
Ellas (Females)	Comer	Com	-en	Comen

The recipe for **-er verbs** is: **remove the -er end of the infinitive** and **add the proper ending**, depending on who or what is performing the action. All conjugations, except for irregular verbs follow this pattern.

Please write the following five times each:

Como:_____

Comes:_____

Come:_____

Comemos:_____

Comeis:_____

Comen:_____

Please write the proper conjugations for the following phrases five times:

I eat: _____.

You eat:_____.

They,He,She eat:_____.

We eat:_____.

They all eat:_____.

They, Males, Females speak:_____.

*You now can conjugate the **present tense of all regular -er Spanish verbs!!***

Find a **regular Spanish verb ending in -er** and write the **six conjugations** of the verb:

_____

_____

_____

The following table is the **Present Tense** conjugation for **ALL REGULAR -ir verbs** in Spanish using the infinitive **Escribir**, meaning to write:

Pronoun	Infinitive	Remove -ir	Add ending	Conjugation
Yo (I)	Escribir	Escrib	-o	Escrib**o**
Tu (You)	Escribir	Escrib	-es	Escrib**es**
Usted (It)	Escribir	Escrib	-e	Escrib**e**
Nosotros (We)	Escribir	Escrib	-emos	Escrib**emos**
Vosotros (They all)	Escribir	Escrib	-eis	Escrib**eis**
Ustedes (They)	Escribir	Escrib	-en	Escrib**en**
Ellos (Males)	Escribir	Escrib	-en	Escrib**en**
Ellas (Females)	Escribir	Escrib	-en	Escrib**en**

The recipe for **-ir verbs** is: **remove the -ir end of the infinitive** and **add the proper ending**, depending on who or what is performing the action. All conjugations, except for irregular verbs follow this pattern.

Please write the following five times each:

Escrib**o**:_____

Escrib**es**:_____

Escrib**e**:_____

Escrib**emos**:_____

Escrib**eis**:_____

Escrib**en**:_____

Please write the proper conjugations for the following phrases five times:

I write: _____.

You write:_____.

They,He,She writes:_____.

We write:_____.

They all write:_____.

They, Males, Females write:_____.

*You now can conjugate the **present tense of all regular -ir Spanish verbs!!***

Find a **regular Spanish verb ending in -ir** and write the **six conjugations** of the verb:

_____

_____

_____

_____

LECCION 3 : PRONUNCACION (Pronunciation)

Spanish is a language known as a "romance language", meaning its origins are Latin in nature. For this reason, spoken Spanish is enunciated without many stresses on consonants like in English. Everyday spoken Spanish is very fluid and there are few abrupt endings. This is why if one tries to speak in Spanish with an English accent, stressing many consonants, the language sounds more abrupt or choppy. The first and only rule of Spanish pronunciation is DON'T STRESS IT.

When speaking/pronouncing a Spanish word or phrase, pretend like you're a magician casting a spell- lightly and flowy, except at the end of the spell. Let's try out our first spell:

English: We are speaking Spanish.

Spanish: Estamos hablando espanol.

Consider how the English sentence is four words, yet the Spanish sentence is only 3 words. **SPANISH IS EASIER THAN ENGLISH BECAUSE YOU CAN SAY A LOT WITH FEWER WORDS.** ☐ This trend continues even into more elaborate Spanish phrases.

English pronounciation: "Oi ahr spee-keng Spahn-ish" -6 Syllables

Spanish pronunciation: "Es-tah-mos ah-blan-do ez-pan-yol." 9 Syllables

Try sounding out the Spanish pronunciation over and over-about 10 times. As you continue repeating the sentence, try to speed up your pronunciation and add the words one right after the other with no pause between words. Remember, **you're a magician casting a spell, not a drill sergeant .**

What do we need to be able to properly pronounce our spells? We have to know the **Spanish alphabet** or **Abecedario**, pronounced "Ah-bay-say-dah-re-oh". For our native English speakers, this section is limited to the variance between the English and Spanish alphabet at the end of the lesson; however, it is recommended that you familiarize yourself with the full Spanish alphabet, **sounding each pronounced letter out ten times.**

A/a- Pronounced "Ah" – **A** as in Apple

B/b-Pronounced "Bey"- **B** as in Baby

C/c-Pronounced "Say"- **C** as in Car

D/d-Pronounced "Day"-**D** as in Dad

E/e-Pronounced "Ey"-**E** as in Everything

F/f-Pronounced "Ehfay" – **F** as in Freedom

G/g- Pronounced "Geh"-**When succeeded by a,o,u G** as in Galvanize, Go, Gullible; **When succeeded by i, e, H** as in Hello.

H/h-Pronounced "Ark-hay"-**Silent** if standalone;

if preceded by "C", CH as in **Ch**arming

I/i-Pronounced "Ee-gree-yay-gah" – **I** as in Interesting

J/j- Pronounced "Hoe-tah"- **H** as in Hello

K/k-Pronounced "Kah" – **K** as in King

L/l-Pronounced "Ehl-lay" – **L** as in Living;

Ll/ll-Pronounced "Ehl-lay doble" -When two in succession, **Y** as in Yes

M/m- Pronounced "Ehm-may" – **M** as in Million

N/n- Pronounced "Ehn-nay" – **N** as in Now

Ñ/ ñ- Pronounced "In-nyay"- **ñ** as in Pequeño "Pay-can-y-yo"

O/o-Pronounced "Oh" – **O** as in Other

P/p-Pronounced "Peh" – **P** as in **P**earl

Q/q-Pronounced "Koo" – **Q** as in **Q**uintet

R/r-Pronounced "Air-ray"-**R** as in **R**ighteous;

Rr/rr-Pronounced "Air-ray dough-blay"- **Trilled R** as in Te**rr**itory

S/s-Pronounced "Essay"- **S** as in **S**uperb

T/t-Pronounced "Tay" -**T** as in **T**errific

U/u-Pronounced "Ooh"-**U** as in **U**nited

V/v-Pronounced "Ooh-bay"- **B** as in **B**ring

W/w-Pronounced "Ooh-bay dough-blay"- **W** as in **W**inning

X/x-Pronounced "Eh-kiz"- **CK**s as in So**ck**s; **X** as in E**x**amination

Y/y-Pronounced "Yay" – **Y** as in **Y**oyo

Z/z-Pronounced "Zeh-tah"-**Z** as in **Z**ebra

Native English Speakers:

G/g- Pronounced "Geh"-When succeeded by **a,o,u G** as in **G**amble, **G**o, **G**ullible; **When succeeded by i, e, H** as in **H**ello.

H/h-Pronounced "Ark-hay"-**Silent** if standalone;

if preceded by "C", CH as in **Ch**arming

L/l-Pronounced "Ehl-lay" – **L** as in **L**iving;

Ll/ll-Pronounced "Ehl-lay doble" -When two in succession, **Y** as in **Y**es

Ñ/ ñ- Pronounced "In-nyay"- **ñ** as in Peque**ñ**o "Pay-ca**ny**-yo"

Rr/rr-Pronounced "Air-ray dough-blay"- **Trilled R** as in Te**rr**itory

*Note to the trilling of the double-r in Spanish**: Allow your tongue to **form "R"** and say **"Really"**. The trick is to allow the air to make the tip of your tongue reverberate at the front of your mouth. If this doesn't work, try pronouncing **"array"** but hold the double r sound and force the air from your throat. Doing this should make the **tip of your tongue reverberate** and make a trilled r sound, even if for a moment. If this doesn't work, **try saying the phrase "Well really" and hold onto the "r", and alternate "r" and "l".***

In conclusion, **you now have all the tools to cast your spells**. The **Abecedario** will be your reference to when you encounter Spanish words that you can't quite speak. Remember Spanish is not stressed with every syllable like English. The sentences you form in Spanish or refer to will be **very fluid**, placing the emphasis only at the end of the sentence. **Congratulations! You are now able to pronounce any word in Spanish!**

LECCION 4: ¿QUE HORA ES? WHAT TIME IS IT?

Telling time is much like telling numbers, and the response to the Spanish question, "Que hora es?" is the same as one would encounter in English. The answer is often **a number between 0 and 24** followed by **a number between 0 and 60**. For simplicity for the reader, **common responses to common Spanish time or number questions are included in this lesson.** Our objective here is **to provide the reader with the tools necessary to respond to persons asking time or number-related questions in Spanish**, beginning with the Spanish numbering system, followed by the Spanish days of the week, followed by the Spanish months of the year, followed by the enumerations of Spanish numbers and their pronunciation.

Numeral	Spanish Word	Pronunciation
1	Uno	"Ooh-noh"
2	Dos	"Doze"
3	Tres	"Trehz"
4	Cuatro	"Kwah-troh"
5	Cinco	"Seen-coh"
6	Seis	"Sayz"
7	Siete	"See-yeh-tay"
8	Ocho	"Oh-cho"
9	Nueve	"New-way-vay"
10	Diez	"Dee-ayz"
11	Once	"Awn-say"
12	Doce	"Doze-say"
13	Trece	"Tray-say"
14	Catorce	"Kah-tore-say"
15	Quince	"Keen-say"

16	Diez y seis	"De-ayz-ee-sayz"
17	Diez y siete	"De-ayz-ee-see-yeh-tay"
18	Diez y ocho	"De ayz-ee-oh-cho"
19	Diez y nueve	"De-ayz-ee-new-wayvay"
20	Veinte	"Ben-tay" **Remember Spanish "V" is pronounced like "B".**
21	Veinte uno	"Ben-tay-ooh-noh"
22	Veinte dos	"Ben-tay-doze"
23-29	Veinte tres, veinte cuatro, veinte cinco, veinte seis, veinte siete, veinte ocho, veinte nueve.	"Ben-tay trehz, kwah-tro, seen-coh, says, see-eh-tay, oh-cho, new-way-vay". **All numbers above 29 follow the pattern**
30	Treinta	"Treh-een-tah"
40	Cuarenta	"Kwah-ren-tah"
50	Cincuenta	"Seen-kwen-tah"
60	Sesenta	"Say-sen-tah"
70	Setenta	"Say-ten-tah"
80	Ochenta	"Oh-chin-tah"
90	Noventa	"Noh-ben-tah"
100	Cien	"See-in"
1000	Mil	"Meel"
10000	Diez mil	"Dee-ayz meel"
100,000	Cien mil	"See-in meel"
1,000,000	Million	"Meel-ee-awhn"
10,000,000	Diez million	"Dee-ayz-meel-ee-awhn"

| 1,000,000,000 | Billion | "Bee-ill-awhn" |
| 1,000,000,000,000 | Trillion | "Tree-ill-awhn" |

Whew! You're doing great, and now you can say numbers from 1 to a trillion.

In Spanish, **when telling time**, specifically the **date**, we say the **number of the day**, the **name of the month**, then the **number of the year** like this:

DAY/MONTH/YEAR

For telling a specific time, we say the hour, minute-and seconds if needed like this:

HOUR:MINUTE:SECOND

The Spanish questions, **"¿Que tiempo es?"**, and **"¿Que hora es?"** have the same meaning: **"What time/hour is it?"**

The Spanish questions, **"¿Que dia es?"**, **"¿Que semana es?"** , **"¿Qué mes es?"** , and **"¿Qué ano es?"** have the meaning, **"What day is it?"**, **"What week is it?"**, **"What month is it?"**, and **"What year is it?"**, respectively. Try practicing these aloud four or **five times.**

Spanish Days of Week:

English Day of Week	Spanish Day of Week	Pronuncation
Monday	Lunes	"Mun-day"/ "Loon-ez"
Tuesday	Martes	"Tooz-day"/ "Mar-tehz"
Wednesday	Miercoles	"Wins-day"/ "Me-air-coal-ez"
Thursday	Jueves	"Theres-day"/ "Way-vez"
Friday	Viernes	"Fry-day" / "Vee-air-nez"
Saturday	Sabado	"Sat-er-day" / "Sah-bah-

		doh"
Sunday	Domingo	"Son-day" / "Doh-meen-go"

Other Time Vocabulary Terms:

Spanish	Pronunciation	Meaning
Dia	"Dee-yah"	Day/Daytime
Manana	"Man-yah-nah"	Morning/Tomorrow
Mediadia	"Meh-dee-ah-dee-yah"	Mid-day/Noontime
Tarde	"Tahr-deh"	Afternoon/Late
Noche	"No-chay"	Evening/Nighttime
Medianoche	"Meh-dee-ah-no-chay"	Mid-night/During night

LECCION 5: "ESTABAMOS PRACTICANDO" WE WERE PRACTICING

The Imperfect Tense of Spanish is used to say what you WERE DOING. **Gerunds**, or words that **end with -ing**, are useful when saying what you WERE DOING. The **Past Participle**, or words that **end in -ed**, are useful when saying what you had already completed in the past. Much like in lessons 1 and 2, the Imperfect Tense is **a matter of creating another spell**. Here's how it works:

For **-ar Verbs**: Remove "ar" and **add the following** depending on the subject:

IMPERFECT TENSE OF -AR SPANISH VERBS

Subject	Verb	Remove "-ar"	Add ending	Conjugation	Meaning
Yo	Hablar	Habl	-aba	Hablaba	I **was speaking**.
Tu	Hablar	Habl	-abas	Hablabas	You **were speaking**.
Usted, El, Ella	Hablar	Habl	-aba	Hablaba	He, she, it **was speaking**.
Nosotros	Hablar	Habl	-abamos	Hablabamos	We **were speaking**.
Vosotros	Hablar	Habl	-abais	Hablabais	They all **were speaking**.
Utds.,Ellos, Ellas	Hablar	Habl	-aban	Hablaban	They **were speaking**.

For **-er** and **-ir** verbs: Remove the **"ir"** or **"er"** and **add the following** depending on the subject:

IMPERFECT TENSE OF -IR SPANISH VERBS

Subject	Verb	Remove "-ir"	Add ending	Conjugation	Meaning
Yo	Escribir	Escrib	- ía	Escrib ía	I **was writing..**
Tu	Escribir	Escrib	- ías	Escrib ías	You **were writing..**
Usted, El, Ella	Escribir	Escrib	-ía	Escrib ía	He, she, it **was writing.**
Nosotros	Escribir	Escrib	- íamos	Escrib íamos	We **were writing.**
Vosotros	Escribir	Escrib	- íais	Esrib íais	They all **were writing.**
Utds.,Ellos, Ellas	Escribir	Escrib	- ían	Escrib ían	They **were writing.**

You can now **practice** each of these conjugations for **any -ar Spanish verb** on the lines below:

_____

_____

_____

_____

IMPERFECT TENSE OF -ER SPANISH VERBS

Subject	Verb	Remove "-er"	Add ending	Conjugation	Meaning
Yo	Encender	Encend	- ía	Encend ía	I **was starting..**
Tu	Encender	Encend	- ías	Encend ías	You **were starting..**
Usted, El, Ella	Encender	Encend	-ía	Encend ía	He, she, it **was starting.**
Nosotros	Encender	Encend	- íamos	Encend íamos	We **were starting.**
Vosotros	Encender	Encend	- íais	Encend íais	They all **were starting.**
Utds.,Ellos, Ellas	Encender	Encend	- ían	Encend ían	They **were starting.**

You can now **practice** each of these conjugations with either **-er** OR **-ir** verbs on the lines below:

_____

_____

_____

_____

_____

_____

_____

GREAT! You've got the hang of how to cast a few more spells!

Now, let's consider words that end in -ing. **Words that end with -ing** are helpful for the past, present, and future tense in Spanish, since you will want to say what you were do**ing** in the past, are do**ing** in the present, and will be do**ing** in the future! Since this is true, we have to know how to change Spanish verbs into their **-ing form**.

___

<u>Here's how: FORMING THE -ING FORM</u>

FOR **-AR SPANISH VERBS**: **Remove** the **-ar** ending, then add **-ando**

Let's use the verb **Hablar**:

Hablar – (ar)+ (ando)= Hablando----Speaking

___

FOR **-ER AND -IR SPANISH VERBS**: **Remove** the **-er** or **-ir** ending, then add **-iendo**

Let's use the verbs **Encender** and **Escribir**:

Encender – (er) +(iendo)=Encendiendo----Starting

Escribir-(er) +(iendo) = Escribiendo----Writing

___

On the lines below, please choose ANY REGULAR -AR SPANISH VERB and form the -ing form of the verb five times:

_____

_____

On the lines below, please choose ANY REGULAR -ER or -IR SPANISH VERB and form the -ing form of the verb five times:

_____

_____

Moving forward, we have to consider **words that end in -ed**, which inform a person that a particular action was **completed in the past**, is actively **being completed**, or **will be completed** at a future time. Since this adds clarity to communicating in Spanish, let's try **forming the -ed form of Spanish verbs**:

Here's how: FORMING THE -ED FORM

FOR -AR SPANISH VERBS: Remove the -ar ending, then add -ado

Let's use the verb **Hablar**:

Hablar – (ar)+ (ado)= Hablado----Spoke

FOR -ER AND -IR SPANISH VERBS: Remove the -er or -ir ending, then add -ido

Let's use the verbs **Encender** and **Escribir**:

Encender – (er) +(ido)=Encendido----Started ; Escribir-(er) +(ido) = Escribido----Wrote

On the lines below, please choose ANY REGULAR -AR SPANISH VERB and form the -ed form of the verb five times:

_____

_____

On the lines below, please choose ANY REGULAR -ER or -IR SPANISH VERB and form the -ed form of the verb five times:

_____

_____

Review: Please say the following:

"Hoy yo practicaba espanol." = Today was practicing Spanish.

"Hoy yo practicando espanol." =Today I am practicing Spanish.

"Hoy yo practicado espanol." = Today I practiced Spanish.

Notice how changing just the verb results in the different meanings? **Conjugations are the key to communicating in Spanish, and this manual provides you with **the formulas for successfully conjugating** and **pronouncing** the Spanish language.

LECCION 6: "PRETERITO PERFECTO" PERFECT PRETERITE

En español, we are able to say what we **have done already in the past**. Recall that the Imperfect Tense with the "-aba" and "-ía" endings are how you say "I,you,he,she,it,we,they all, they" was do**ing**. The Preterito Perfecto is how we tell something **has already been done**. The difference in the Imperfect Tense and the Preterito Perfecto is that the Preterito Perfecto utilizes the Imperfect Tense as one of its ingredients to ensure that one knows the action was in the past. Put scientifically, **the Preterito Perfecto is mutually inclusive of the Imperfect Tense.**

Here's how we form the Preterito Perfecto:

1. Write the **Imperfect Tense** of the verb **HABER**-meaning **TO HAVE**. Since you are able to conjugate the Imperfect Tense, you know we have to **drop the -er ending** and **add – ía:**

Yo	Hab **ía**	Nosotros	Hab **íamos**
Tu	Hab **ías**	Vosotros	Hab **íais**
Usted,El,Ella	Hab **ía**	Uds.,Ellos, Ellas	Hab **ían**

2. Write the **PAST PARTICIPLE**, or the -ed form of the action that was completed, recalling that to form the **-ed form** of -ar verbs, we **remove the -ar** and **add -ado**; for **-er** or **-ir** verbs, we remove the **-er** or **-ir** and add **-ido:**

Hablar	-(ar)+ (ado)	Habl**ado**
Encender	-(er) + (ido)	Encend**ido**
Escribir	-(ir) + (ido)	Escrib**ido**

That's all there is to it. Putting all of our ingredients together, we come out with the following **very useful combinations of phrases** used in **everyday Spanish:**

Let's use **HABLAR** to say what the subject had completed (**speaking**) in the past.

HABLAR "TO SPEAK"	
Yo hab **ía hablado.**	Nosotros hab **íamos hablado.**
Tu hab **ías hablado.**	Vosotros hab **íais hablado.**
Ud. Ello, Ella hab **ía hablado.**	Uds.,Ellos, Ellas hab **ían hablado.**

I **had spoken.**	We **had spoken.**
You **had spoken.**	They all **had spoken.**
It, She, He **had spoken.**	They, Girls, Boys **had spoken.**

Let's use **ENCENDER** to say what the subject had completed (**starting**) in the past.

ENCENDER "TO START/BEGIN"	
Yo hab **ía encendido.**	Nosotros hab **íamos encendido.**
Tu hab **ías encindido.**	Vosotros hab **íais encendido.**
Ud. Ello, Ella hab **ía encendido.**	Uds.,Ellos, Ellas hab **ían encendido.**

I **had started.**	We **had started.**
You **had started.**	They all **had started.**
It, She, He **had started.**	They, Girls, Boys **had started.**

Let's use **ESCRIBIR** to say what the subject had completed (**writing**) in the past:

ESCRIBIR "TO WRITE"	
Yo hab **ía escribido.**	Nosotros hab **íamos escribido.**
Tu hab **ías escribido.**	Vosotros hab **íais escribido.**
Ud. Ello, Ella hab **ía escribido.**	Uds.,Ellos, Ellas hab **ían escribido.**

I **had written.**	We **had written.**
You **had written.**	They all **had written.**
It, She, He **had written.**	They, Girls, Boys **had written.**

NOW IT'S YOUR TURN!

Practice conjugating the **(6) IMPERFECT TENSE** conjugations of **HABER** in the table below:

Yo		Nosotros	
Tu		Vosotros	
Usted, El, Ella		Uds. Ellos, Ellas	

Now, select **ANY REGULAR SPANISH VERB** and **write the -ed form** on the line below:

_____

Finally, **add the conjugations of HABER** and the **-ed form** on the lines below, writing the **phrase five times:**

_____

_____

_____

_____.

LECCION 7: "¿QUE ESTA TRABAJANDO MANANA?" WHAT ARE YOU WORKING TOMORROW?

En espanol, we can say what we will be doing in the future. Forming the FUTURE TENSE is as simple as adding the appropriate ending to the action word. This spell requires no removal, **only addition**!

Here's how:

Let's use the verb **HABLAR**:

Subject	Add ending	Conjugation	Meaning
Yo	**-é**	Hablar **é**	I **will speak.**
Tu	**-ás**	Hablar**ás**	You **will speak.**
Usted, El, Ella	**-á**	Hablar **á**	It,She,He **will speak.**
Nosotros	**-emos**	Hablar **emos**	We **will speak.**
Vosotros	**-éis**	Hablar **éis**	They all **will speak.**
Uds. Ellos, Ellas	**-án**	Hablar **án**	Those, boys, girls **will speak.**

Pick any **REGULAR -AR Spanish verb** and write the (6) **FUTURE TENSE conjugations** on the lines below:

_____

_____

_____

_____

_____.

Let's do the same with our verbs **ENCENDER** and **ESCRIBIR**:

Subject	Add ending	Conjugation	Meaning
Yo	-é	Encender **é**	I **will start.**
Tu	-ás	Encender**ás**	You **will start.**
Usted, El, Ella	-á	Encender **á**	It,She,He **will start.**
Nosotros	-emos	Encender **emos**	We **will start.**
Vosotros	-éis	Encender **éis**	They all **will start.**
Uds. Ellos, Ellas	-án	Encender **án**	Those, boys, girls **will start.**

Subject	Add ending	Conjugation	Meaning
Yo	-é	Escribir **é**	I **will write.**
Tu	-ás	Escribir**ás**	You **will write.**
Usted, El, Ella	-á	Escribir **á**	It,She,He **will write.**
Nosotros	-emos	Escribir **emos**	We **will write.**
Vosotros	-éis	Escribr **éis**	They all **will write.**
Uds. Ellos, Ellas	-án	Escribir **án**	Those, boys, girls **will write.**

Now that you've seen the **Future Tense** in action, let's see **some examples**:

Hablar**é** con ella.-----I will speak to her.

Comeremos en casa de Jose. ----- We will eat at Jose's.

Lara no volver**á**.-----Lara won't come back.

¿Lo entender**ás?** -----Will you understand it?

¿Me quier**ás** esperar un momento, por favor?-----Will you wait for me a momento, please?

Me levantaré temprano. ----I'll get up early.

In addition to the **FUTURE TENSE**, there is an **informal future tense** that is formed using the **IRREGULAR VERB "Ir" which means "To-go"**. This is useful when you are going to be **saying where you are going** and **what you are going to do**. This **informal future tense** usually includes **a destination** or **an activity** as the **object of the sentence**. This is what **makes the verb "Ir" irregular: It's regular conjugation remains the same in the present and future tense.**

Here's how:

Yo	**Voy**= I go.	Nosotros	**Vamos**=We go.
Tu	**Vas**=You go.	Vosotros	**Vais**=They all go.
Ud. El, Ella	**Va**=He,she, it goes.	Uds. Ellos, Ellas	**Van**=They, boys, girls, go.

Here are some examples:

Yo **voy a hablar** con ellos. -----I am going to speak with the boys.

Tu **vas a encender** el proyecto en martes.---You are going to begin the project on Tuesday.

¿Usted **va al resturante** favorita de Miguel?---Are you going to Miguel's favorite restaurant?

El **va al zoo** para buscar los animales.----He is going to the zoo to look at the animals.

Ella **va al biblioteca** en fin de semana.----She is going to the library on the weekend.

Nosotros **vamos al pelicula** en jueves.---We are going to the movie on Thursday.

Ustedes **van al tournament** de futbol en Nashville. ---You all are going to the soccer tournament in Nashville.

Ellos **van al iglesia** en domingo.----The boys go to church on Sunday.

Ellas **van al casa** de Marta en miércoles. ----The girls go to Marta's house on Wednesday.

LECCION 8: "!PRACTICAS VOCABULARIO Y BUSCANDO EL MUNDO DIFERENTE!" PRACTICE VOCABULARY AND SEE THE WORLD DIFFERENTLY!

This is the final lesson of Spanish in this manual, and these INFINITIVES and WORDS will enable you to say what you would like in the past, present, and future. We have included a section dedicated to the medical profession as well as the pharmaceutical industry as a cross-reference for both Spanish and English speakers alike. You have your manual to fall back on, and you can do anything you set your mind to. Let's have a look at what you've done so far:

Lección 1: CONJUGATIONS OF REGULAR VERBS

Lección 2: CONJUGATIONS PRACTICE AND PRONOUNS

Lección 3: PRONUNCACION

Lección 4: ¿QUE TIEMPO ES? WHAT TIME IS IT?

Lección 5: "ESTABAMOS PRACTICANDO" WE WERE PRACTICING

Lección 6: "PRETERITO PERFECTO" PERFECT PRETERITE

Lección 7: "¿QUE ESTA TRABAJANDO MANANA?" WHAT ARE YOU WORKING TOMORROW?

Consider this your grocery list or materials list for your spells as you go forth and cast your Spanish forth into the world. Each of these words and verbs are commonly used in the Spanish language socially and professionally.

We have compiled this comprehensive vocabulary list as well as a glossary of common pharmacological terms as a practical reference for the socialite and professional who wishes to communicate effectively in their relationships with Spanish speakers as well as Spanish speakers who wish to learn English.

We encourage you to read this list, select words that you believe to be important in your communications, and practice each by writing them on the work pad at the end of this manual. We hope you see the world differently.

VOCABULARIO REGULAR Y SOCIAL/ REGULAR AND SOCIAL VOCABULARY

"Doctor" = Medico	"Screen"=Ventana
"Water"=Aqua	"Keys"=Lleves
"Food"= Comida	"Email" = Correo electronico
"Sad" = Triste	"Communicate"=Communicar
"Mad"=Enojado	"Establish"=Estabilir
"Hot"=Caliente	"Present/Show"=Presentar
"Cold"=Frio	"To give"= Dar
"Building" =Edificio	"To sleep"=Descansar
"Library"=Biblioteca	"To feel"=Tacitar
"Seeing"=Buscando	"To hear"=Oir
"Believing"=Creyendo	"To learn"=Aprender
"Thinking"=Pensando	"To live"=Vivir
"Living"=Viviendo	"To write"=Escribir
"Cooking"=Cocinando	"To listen"=Escuchar
"Room"=Salon	"To operate"=Manejar
"Office"=Oficina	"To forget"=Olvidar
"Chair"=Sillo	"To remember"=Recojar
"Light"=Luz	"To encounter"=Encontrar
"Dark"=Oscuro	"To improve"=Improbar
"Shy"=Timido	"To decline"=Declinar
"Clothes"=Quedas	"To affirm"=Afirmar
"Pants"= Pantalones	"To open"=Abrir
"Socks"=Calcetienes	"To enter"=Entrar
"Folder"=Carpeta	"To exit"=Exitar

"Side"=Lado	"To smoke"=Fumar
"Up"=Arriba	"To be funny"=Comicar
"Down"=Abajo	"To be beautiful"=Bonito/Bonita
"Left"=Izquierda	"To be appreciative"=Apreciar
"Right"=Derecha	"To be happy"=Disfrutar
"Speaking"=Hablando	"To chose"=Seleccionar
"Writing"=Escribiendo	"To decide"=Decidir
"Playing"=Tocando	"To discipline"=Disciplinar
"To call/to name"=Llamar	"To carry,bring"=Llevar
"To owe"=Deber	"To stay,remain"=Quedar
"To wait for/hope for"=Esperar	"To exist"=Existir
"To work"=Trabajar	"To receive, welcome, greet"=Recibir
"To change"=Cambiar	"To need,require"=Necesitar
"To form, shape, fashion"=Formar	"To divide, leave"=Partir
"To accept, approve"=Aceptar	"To get, obtain, achieve"=Lograr
"To study"=Estudiar	"To please, be pleasing"=Gustar
"To help"=Ayudar	"To fulfill, carry out"=Cumplir
"To try, attempt"=Intentar	"To raise, to life"=Levantar
"To ask, inquire"=Preguntar	"To register"=Abanderar
"To go to bed"=Acostar	"To love"=Amar
"To prepare"=Preparar	"To laugh"=Reir
"To go out"=Salir	"To translate"=Traducir
"To sale"=Vender	"To fly"=Volar
"To train"=Adiestrar	"To master"=Para dominar

Vocabulario de Nautica / Nautical Vocabulary

Abaft – "Hacia popa"	Abeam – "Por el través"
Aboard – "A bordo"	Adrift – "A la deriva"
Advection fog – "Niebla de advección"	Aft – "A popa"
Aground – "Encallado"	Ahead – "Avante, adelante"
Aids to navigation (ATON) – "Senzlicazion meritima; ayudas a la navegación"	Air draft – "Corriente de aire"
Air exhaust – "Purga de aire"	Air intake – "Toma de aire"
Allision – "Abordaje"	Aloft – "Arriba, en lo alto"
Alternator – "Alternador"	Amidships – "En el centro del buque; en curjia"
Anchor – "Ancla"	Anchor bend (fisherman's bend) – "Cote de ancla (nudo de pescador)"
Anchor light – "Luz de fondeo"	Anchor rode – "Cabo/cadena del ancla"
Anchor well – "Pozo de anclas"	Anchor's aweigh – "Levar anclas"

Anchorage area – "Fondeadero"	Aneroid barometer – "Barometro aneroide"
Apparent wind – "Viento aparente"	Astern – "A popa"
Athwartship – "De banda a banda"	Attitude – "Aspecto/posicion"
Automatic pilot – "Piloto automatico"	Auxiliary engine – "Motor auxiliar"
Back and fill – "Fachear y marear en viento"	Backing plate – "Placa de soporte"
Backing spring (line) – "Cabo para remolque en reversa"	Backstay – "Burda/ estay de popa"
Ballast – "Lastre"	Bar – "Barra"
Barge – "Barcaza"	Barograph – "Barografo"
Barometer – "Barometro"	Bathing Ladder – "Escala de bano"
Batten – "Cerrar"	Batten down! – "Cerrar las escotillas"
Batten pocket – "Vaina de sable"	Battery – "Bateria"
Battery charger – "Cargador de batería"	Beacon – "Baliza"
Beaker line – "Linea de rompientes"	Beam – "Manga"

Beam reach – "Navegar de través"	Bear off – "Desatracar/alejarse"
Bearing – "Marcacion; demora"	Beating – "Navegar de bolina"
Beaufort Wind Scale – "Escala Beaufort de vientos"	Before the wind – "Navegar con viento en popa"
Bell buoy – "Boya con campana"	Below – "Bajo cubierta"
Belt – "Correa (de transmisión)"	Berth – "Amarradero, atracadero"
Bilge – "Sentina"	Bilge alarm system – "Alarma de sentina"
Bilge drain – "Bomba de sentina"	Bimini top – "Toldo Bimini"
Binnacle – "Bitacora"	Binoculars – "Prismaticos"
Bitt – "Bita"	Bitter end – "Chicote"
Block – "Moton"	Boarding Ladder – "Escala real"
Boat hook – "Bichero"	Bollard – "Bolardo"
Bolt rope – "Relinga (de vela)"	Boom – "Botavara"
Boom vang (rigid) – "Contra de botavara (rígido)"	Bosun's chair – "Guindola de arboladura"

Bottlescrew – "Tensor"	Boundary layer – "Agua del timón"
Bow – "Proa"	Bow (lateral) – Amura
Bow fitting – "Herraje de proa"	Bow line – "Largo de proa"
Bow truster – "Helice de proa"	Bowsprit – "Baupres"
Braided rope – "Cabo trenzado"	Breakaway – "Interrumpir la maniobra"
Breaker – "Rompiente (ola)"	Breaking strength (BS) – "Resistencia a la rupture"
Breakwater – "Reompeolas"	Breast line – "Amarra de costado"
Bridge (on a ship) – "Puente de mando"	Bridge markings – "Senalizacion de puentes"
Bridle – "Yugo (en el extremo del cabo de remolque)"	Broach – "Hacer capilla"
Broad on the beam – "Por el traves"	Broad reach – "Navegar a un largo"
Broadcast – "Trasmision por radio"	Broadcast (to) – "Transmitir"
Broadcast notice to mariners – "Radioaviso a los navegantes"	Broadside to the sea – "Olas por el través"
Bulkhead – "Mamparo"	Bullnose (hawse hole) – "Escoben"

Bunk – "Litera"	Buoy – "Boya"
Buoy moorings – "Boyas de amarre"	Buoy station – "Posicion asignada a la boya"
Buoyage – "Sistema de boyado"	Buoyancy – "Flotabilidad"
Cabin – "Cabina"	Call sign – "Signo de llamada"
Cam cleat – "Galapago"	Can (buoy) – "Boya cilíndrica"
Capsize – "Zozobrar"	Car – "Carro"
Carburator – "Carburador"	Cardinal marks – "Puntos/marcas de boyado marítimo"
Cast iron – "Hierro fundido"	Cast off – "Zarpar"
Catamaran – "Catamaran"	Catenary – "Catenaria"
Cavitation – "Cavitacion"	Celestial navigation – "Navegación astronómica"
Center of gravity – "Centro de gravedad"	Center point method, rectangular area, bearing and distance (SAR) – "Buscar dentro de una zona rectangular a partir del centro con ruta y distancia pedeterminadas (método SAR)"
Centerline – "Crujia"	Chafe – "Desgastarse por el roce"

Chafing gear – "Material de protección (para cabos)"	Chain locker – "Caja de cadenas"
Chainplate – "Plancha de cubierta"	Change oil – "Cambiar el aceite"
Channel (nav.) – "Canal"	Channel (radio) – "Canal"
Characteristic (ATON) – "Características del boyado"	Chart – "Carta nautica"
Chart reader – "Lector de cartas"	Chart table – "Mesa de cartas"
Chine – "Quiebro"	Chock – "Guia"
Chop – "Mar picada"	Cleat – "Cornamusa"
Clevis pin – "Perno de fijación"	Clew – "Puno de escota"
Close reach – "Navegar de cenida abierta"	Close-hauled – "Navegar de bolina"
Closeout – "Convergencia de (olas) rompientes"	Closing – "Aproximarse a"
Clove hitch – "Ballestrinque"	Clutch – "Embrague"
Coach roof – "Carroza"	Coaming – "Brazola"
Coast Guard- approved – "Aprobado por la Guardia Costera"	Cockpit – "Banera"

Coil down – "Adujar"	Cold front – "Frente frio"
Colors – "Bandera nacional"	Comber – "Ola a punto de romper"
Combination buoy – "Boya con senales luminosas y acústicas"	Combustion – "Combustion"
Companionway – "Tambucho"	Companionway hatch – "Escotilla del tambucho"
Compartment – "Compartimento"	Compass – "Compas"
Connection – "Toma de corriente exterior"	Control Signal Light Gun – "Pistola de luces de control"
Cooling (wáter, air) – "Enfriemiento (del aqua, del aire)"	Conventional direction of buoyage – "Direccion convencional de balizamiento"
Corner method (SAR) – "Búsqueda dentro de una área definida por ángulos (método SAR)"	Coordinate – "Coordinar"
Cotter pin – "Pasador de chaveta"	COSPAS-SARSAT System – "Sistema COSPAS-SARSAT"
Cove – "Ensenada"	Course – "Rumbo"
Cowls – "Manguerote de ventilación"	Coverage factor (C) – "Factor de cobertura (SAR)"
Crab – "Derrapar"	Coxswain – "Patron del bote"
Crash stop – "Parada repentina del motor"	Craft – "Embacacion"

Cringle – "Grillete de vela"	Crest – "Cresta"
Crucifix – "Columna cruciform (p.e. la bita Samson)"	Crossing situation – "Situación de cruce"
Cutter (type of sailboat) – "Cuter"	Current (ocean) – "Corriente (océano)"
Cylinder head – "Cabeza del cilindro"	Cylinder – "Cilindro"
Datum – "Datum"	Damage control – "Control de danos"
Daybeacon – "Baliza diurna"	Davit – "Pescante para dingby"
Dayshape – "Marca diurna"	Dayboard – "Marca diurna"
Dead reckoning – "Navegación a la estima"	Dead in the wáter – "Sin arrancada"
Deck – "Cubierta"	Deadman's stick (static discharge wand) – "Varita para descargar, electricdad estatica"
Deck scuttle – "Escotilla de acceso"	Deck fitting – "Herraje de cubierta"
Deep "V" hull – "Carena en "V"	Deck stepped mast – "Mastil con fogonadura en la cubierta"
Desmoking – "Ventilación"	Depth Finder (sounder) – Escandallo (sondador)
Deviation – "Desviación magnética"	Destroyer turn – "Virada de emergencia"

Diesel (fuel) – "Gasoleo (combustible)"	De-watering – "Achque, achicar"
Digital selective calling (DSC) – "Llamada digital selectiva (CSN)"	Diesel engine – "Motor diesel"
Direction of current –"dirección de la corriente"	Dinghy – "Chinchorro"
Direction of winds – "Direccion del viento"	Direction of waves – "Dirección de la olas"
Displacement – "Desplazamiento"	Dimasting – "Desarbolar"
Distress – "Peligro"	Displacement hull – "Casco de desplazamiento"
Ditching – "Amaraje"	Distress beacon – "Baliza de emergencia"
Dock – "Atracadero"	Do you read me? – "Me recive?"
Dodger – "Capota antirociones"	Dock (to) – "Atracar"
Douse – "Arriar (velas)"	Dolphin – "Duque de alba"
Downwind – "A sotavento"	Downwash – "Compresion del aire causada por el rotor de un helicóptero"
Drag – "Resistencia hidrodinámica"	Draft (draught) – "Calado"
Drift (due to wind) – "Abatimiento"	Drift (due to current) – "Velocidad"

Drop pump – "Bomba de emergencia"	Drogue – "Ancla flotante"
Duct – "Tubo de ventilación"	Dry pump – "Traje estanco"
Ease – "Lascar"	Dynamic forces – "Fuerzas dinámicas"
Ebb current – "Corriente de reflujo"	Ebb – "Marea de reflujo"
Eddy – "Remolino (de viento o de agua)"	Ebb direction – "Direccion de reflujo"
Electrical panel – "Cuadro eléctrico"	Eductor – "Eyector"
Emergency signal mirror – "Espejo de señalización"	Electronic navigation – "Navegación electrónica"
Engine block – "Bloque del motor"	Engine battery – "Bateria del motor"
Engine emissions – "Gases de escape del motor"	Engine control panel – "Cuadro de mandos del motor"
Engine motor – "Maquina"	Engine filter – "Filtro del motor"
Engine starter – "Motor de arranque"	Engine power – "Potencia motriz"
Estimated position – "Fuerzas ambientales"	Environmental forces – "Fuerzas ambientales"
European Union certification – "Certificacion UE"	Estimated position – "Posicion estimada"

Eye of the wind – "Filo del viento"	Eye – "Gaza"
Fairlead – "Guia"	Eye splice – "Gaza tenzada"
Fake down – "Adujar, coger de redondo"	Fairways (mid-channel) – "Canal navegable"
Fast(ening) – "Amarrar, hacer firme"	Fall off – "Caer a sotavento"
Fender – "Defensa"	Fatigue – "Fatiga"
Ferry – "Transbordador"	Fender board – "Defensas montadas a una plancha horizontal"
Fiberglass – "Fibra de vidrio"	Fetch – "Fetch"
Figure eight know – "Nudo de saboya, nudo en ocho"	Fid – "Pasador (para abrir los cordones de un cabo)"
Fitting – "Herraje"	Fill up (fuel) – "Repostar (combustible)"
Fixed light – "Luz fija"	Fix – "Marcacion"
Flare (hull) – "Abanico"	Flame arrester – "Parallamas"
Flash – "Destello"	Flare (light) – "Bengala"
Flemish (down) – "Adujar"	Flashing light – "Luz de destellos"

Flood – "Marea creciente"	Floating aid to navigation – "Senal maritima flotante"
Flood direction – "Direccion de la marea creciente"	Flood current – "Agua de creciente"
Fluke – "Una (del ancla)"	Floor (hull) – "Varenga (del casco)"
Folding – "Hélice con palas abatibles propeller"	Foam crest – "Cresta de espuma"
Fore – "De proa"	Foot (of a sail) – "Pujamen (de la vela)"
Foredeck – "Cubierta de proa"	Fore and aft – "De proa a popa"
Forward – "Hacia proa"	Forestay – "Estay de proa"
Founder – "Hundirse"	Foul – "Encepado (ancla) ; abromado (casco)"
Frames – "Cuadernas del casco"	Four-stroke engine – "Motor de cuatro tiempos"
Freeboard – "Francobordo"	Free communication with the sea – "Embarcar aqua"
Fuel capacity – "Capacidad de combustible"	Front (meteo.) – "Frente"
Fully battened (sail) – "Vela de sables forzados"	Fuel gauge – "Indicador del nivel de combustible"
Funnel (ship) – "Chimenea"	Funnel (fuel) – "Embudo"

Gaff rig – "Aparejo de cangreja"	Furl (to) – "Enrollar, aferrar (sail)"
Gas locker – "Panol para el gas"	Galley – "Cocina"
Genoa – "Genoa"	Gasoline engine – "Motor de gasolina"
Give way vessel – "Embarcación que cede el paso"	Gimbals – "Balancín"
Gooseneck – "Gansera"	Gong buoy – "Boya gong"
Grabline – "Guirnalda salvavidas"	Grab rail – "Pasamano"
Grommet – "Ollao"	Greenwich mean time (GMT) – "Hora del meridian Greenwich"
Group-flashing light – "Luz con grupo de destellos"	Ground fog – "Niebla de radiación"
Gunwale – "Regala"	Group-occulting light – "Luz con grupo de ocultaciones"
Halyard – "Driza"	Half hitch – "Cote"
Hand lead – "Sonda de mano"	Hand bearing compass – "Compas de marcar"
Hanging locker – "Taquilla, panol"	Handrail – "Pasamano"
Hank on (to) – "Engarruchar (una vela)"	Hank (jib) – "Anillo (foque)"

Harness – "Arnés"	Harbor – "Puerto"
Hatch cover – "Tapa de la escotilla"	Hatch – "Escotilla"
Hawser – "Estacha"	Hawsepipe – "Bocina de escoben"
Head (of a sail) – "Puno de driza"	Head – "Bano (retrete)"
Heading – "Rumbo"	Head up! (heads up) – "Orzar a la banda (!Atencion!)
Heave to – "Pairar, fachear"	Headway – "Arrancada"
Heave! (an object) – "!Halar!"	Heave! (a line) – "!Tesar!"
Heavy weather – "Mar gruesa y viento fuerte"	Heaving line – "Guia"
Helm – "Sistema de mando del timón"	Heel – "Escora"
High tide – "Pleamar"	High seas – "En alta mar"
Hoist – "Izar, arriar"	Hitch – "Vuelta de cabo"
Holding tank – "Deposito de aquas negras"	Hoisting cable – "Cabo de salvamento"
Horshoe buoy – "Aro salvavidas tip herradura"	Hole – "Agujero"

Hull – "Casco"	House battery – "Batería auxiliar"
Hull-deck joint – "Junta entre casco y cubierta"	Hull integrity – "Integridad del casco"
Hypthermia – "Hipotermia"	Hurricane – "Huracán"
Impeller – "Impulsor"	I spell – "Deletreo"
In step – "Sincronizado"	In Irons – "Al pairo"
Incident Command System (ICS) – "Sistema de gestionar situaciones de emergencia nacional (ICS)"	Inboard – "Dentro del buque"
Information marks – "Señales marítimas de información"	Inflatable – "Inflabale (chinchorro)"
Injector – "Inyector"	Injection pump – "Bomba de inyección"
In-mast furling – "Enrollador en el mástil"	Inlet – "Ensenada, abertura"
Inverter (electr.) – "Inversor"	Interface – "Zona entre superficies de contacto"
Isobars – "Isobarra"	Inverter (mech.) – "Inversor"
Jackline – "Nervio"	Isolated danger mark – "Señal de peligro aislado"

Jam cleat – "Mordaza:	Jacob's Ladder – "Escala de viento"
Jetty – "Escollera"	Jammer – "Mordaza"
Junction – "Bifurcación de canal"	Jib – "Foque"
Kapok – "Capoc"	Junction aid (obstruction aid) – "Marca 'bifurcación, canal preferido' (marca de peligro aislado)"
Ketch – "Queche"	Keel – "Quilla"
Knot (kn or kt) – "Nudo"	Kicker hook (skiff hook) – "Bichero para enganchar el remolque"
Landmark – "Punto de referencia en tierra"	Knotmeter – "Log corredera"
Lateral marks – "Marcas laterales"	Landmark boundaries method (SAR) – "Delimitacion del area de busqueda con punto referencia en tierra (método SAR)"
Lateral system of buoyage – "Sistema lateral de boyado maritimo"	Lateral system – "Sistema lateral"
Lazarette – "Lazareto"	Latitude – "Latitud"
Leech – "Balumen"	Lee helmed (boat) – "Barco propenso a abatir hacia sotavento"
Leeway (due to current) – "Deriva"	Leeward – "A sotavento"

Length on the water line (LWL) – "Linea de flotacion"	Leeway (due to wind) – "Abatimiento"
Life jacket – "Chaqueta salvavidas"	Length overall (LOA) – "Eslora total"
Life ring (ring buoy) – "Aro salvavidas"	Life raft – "Almadia"
Light buoy – "Boya luminosa"	Lifeline – "Guardamancebos"
Light rhythms – "Ritmos de las luces"	Light List – "Lista de faros y señales"
Lighthouse – "Faro"	Light sector – "Sector de las luces"
Limber holes – "Imbornal de varenga"	Lightning protection – "Protección contra rayos"
Linestopper – "Mordaza"	Line – "Cabo"
Local Notice to Mariners – "Aviso a los navegantes"	List – "Escora"
Log – "Corredera"	Locker – "Panol"
Long splice – "Ayuste largo"	Logbook – "Libro de bitacora"
Longitudinal – "Longitudinal"	Longitude – "Longitud"
Lookout – "Vigia, serviola"	Longshore current – "Corriente paralela a la costa"

Loud and clear – "Fuerte y claro"	LORAN-C – "LORAN-C"
Low battery alarm – "Alarma de bateria baja"	Loud hailer – "Altoparlante"
Luff up- "Orzar"	Lubber line – "Linea de fe luff gratil (vela)"
Magnetic course – "Rumbo magnetico"	Magnetic compass – "Compas magnetico"
Man over board – "Hombre al aqua"	Mainsail – "Vela mayor"
Marine Assistance Broadcast (MARB) – "Radiodifusión de las llamadas de socorro"	Marina – "Marina"
Maritime – "Martitimo"	Marine sanitation device (MSD) – "Retrete marino"
Marline – "Merlin"	Mark – "Senal"
Marlinspike seamanship – "Arte de marinería"	Marlinspike – "Pasador, burel"
Mast head – "Tope del palo"	Mast – "Palo"
MAYDAY – "MAYDAY"	Masthead light – "Luz de tope"
Meridian – "Meridiano"	Medical evacutation (MEDEVAC) – "Evacuacion medica (MEDEVAC)"
Microwave oven – "Horno microondas"	Messenger – "Cabo mensajero"

Mizzen mast – "Palo de mesana"	Mid-channel – "En el medio del canal"
Mooring buoy – "Boya de amarre"	Modified U.S. Aid System – "Sistema de Balizamiento de EEUU (modificado)"
Motor Lifeboat (MLB) – "Embarcacion de salvamento motorizada"	Mooring line- "Cabo de amarre"
Mousing – "Trinca de gancho"	Motorsailer – "Motovelero"
Nautical chart- "Carta nautica"	Nautical Almanac – "Almanaque nautico"
Nautical slide rule – "Regla de calculo tipo náutico"	Nautical mile – "Milla náutica"
Navigable Waters – "Aquas navegables"	Navigable channel – "Canal navegable"
Navigation aids/light (vessels) – "Luz"	Navigation – "Navegación"
Navigation lights – "Instrumentos de navegación"	Navigation instruments – "Instrumentos de navegación"
Navigation Rules (Colregs) – "Reglamento internacional para prevenir los abordajes en la mar"	Navigation lights – "Lucas de situación"
Neap tide – "Marea de cuadratura"	N-Dura hose – "Manguera N-Dura para incendios"
Noise – "Ruido"	Night sun – "Faro de iluminación del helicóptero"

Notice to Mariners – "Aviso a los navegantes"	Normal endurance – "Limite normal de resistencia"
Oars – "Remos"	Nun buoy (conical) – "Boya cónica"
Officer of the Deck (OOD) – "Oficial de guardia"	Occulting light – "Luz ocultante"
Oil pump – "Bomba de aceite"	Offshore – "En alta mar"
On scene – "En la zona de los hechos"	Oil rig – "Plataforma marina
Opening – "Rumbo divergente"	On Scene Commander (OSC) – "Comandante en escena"
Out – "Terminado"	Out of step – "Fuera de fase"
Outdrive – "Unidad inferior"	Outboard – "Fuera de borda"
Overdue – "Retresado"	Over – "Cambio"
Overhead – "Techo (cielo)"	Overhauling the fire – "Dispersar el residuo del incendio"
Pacing – "Sincronizados"	Overload – "Sobrecargar"
Painter line – "Boza"	Pad eye – "Cancamo"
Parallax error – "Error de paralaje"	PAN PAN, PAN PAN, PAN PAN – "PAN PAN, PAN PAN, PAN PAN"

Parallel approach – "Abordar en paralelo"	Parallel – "Paralelo"
Parallel track pattern (SAR) – "Busqueda por rutas paralelas (SAR)"	Parallel rule(rs) – "Reglas paralelas"
Pay out (a line) – "Filar/dar linea"	Passenger space – "Compartimiento para pasajeros"
Pelorus – "Pinula"	Pedestal – "Pedestal"
Personal marker light (PML) – "Luz para chaleco salvavidas"	Pennant – "Gallardete"
Pier – "Muelle"	Persons on board (POB) - "Personas a bordo"
Pilot – "Practico"	Piling – "Estaca"
Pitchpole – "Irse por ojo"	Pitch – "Cabecear"
Planking – "Tablazón"	Planing hull – "Casco de planeo"
Point (to) – "Cenir el viento"	Plimsoll mark – "Marca de francobordo (marca Plimsoll)"
Port – "A babor"	Polyethylene float line – "Cabo flotante de polietileno"
Port tack – "Amurado a babor"	Port light – "Ojo de buey"
Preferred channel mark – "Senal de canal principal"	Power boat – "Bote de motor"

Primary aid to navigation – "Ayudas a la navegación primarias"	Preventer (line) – "Freno de botavara (cabo)"
Probability of detection (POD) – "Probabilidad de avistar (en la operación SAR)"	Prime meridian – "Primer meridiano"
Proceeding from seaward – "Regresando de la mar"	Probability of success (POS) – "Probabilidad de exito (en la operación SAR)"
Propeller – "Hélice"	Prop wash – "Estela de la hélice (turbulencia)"
Propeller shaft – "Eje de cola"	Propeller (two-blade, three-blade) – "Hélice de dos palas, de tres palas"
Protractor – "Transportador de ángulos"	Propeller stuffing box (stern gland) – "Prenaestopas del eje de cola"
Pump – "Bomba (de achique)"	Pulpit (bow) – "Pulpito de proa"
Pyrotechnics – "Material pirotecnico"	Pushpits – "Pulpito de popa"
Quarter – "Aleta"	Quarantine anchorage buoy – "Señal de fondeadero de cuarentena"
Radar beacon (RACON) – "Respondedor radar (RACON)"	RADAR – "RADAR"
Radiation fog – "Niebla de radiacion"	Radar reflector – "Reflector de radar"
Radio direction Finder – "Radioniometro"	Radio contact – "Contacto por radio"

Radio distress call (PAN) – "Llamada de urgencia"	Radio distress call (MAYDAY) – "llamada de socorro"
Radio frequency – "Frecuencia de radio"	Radio distress call (SECURITE) – "Llamada de seguridad"
Radio station – "Canal de radio"	Radio silence – "Silencio de radio"
Radiobeacon – "Radiofaro"	Radio watch – "Radioescucha"
Radome – "Cupula protectora del radar"	Radionavigation – "Radionavegacion"
Range – "Alcance; enfilación"	Rake – "Inclinación"
Red, right, returning – "Regresando de la mar, dejar a la derecha la señal roja"	Range lights – "Señales luminosas de enfilación"
Reef knot – "Nudo llano"	Reef – "Arrecife"
Reefing lines – "Matafiones"	Reefing (a sail) – "Tomar rizos"
Re-flash watch – "Servicio de guardia para evitar nuevos focos de incendio"	Reefing points – "Hilera de ollaos"
Regulatory marks – "Señales reguladoras"	Re-float – "Volver a poner al flote una embarcación hundida"
Rescue swimmer – "Nadador rescatador"	Rescue basket – "Cesta de rescate flotante"
Rhumb line – "Linea loxodrómica"	Retroflective material – "Material retroflectivo"

Rigid inflatable boat (RIB) – "Embarcación rígida inflable (RIB)"	Rig – "Aparejar; instalar; aparejo"
Riprap – "Escollera de defensa"	Rip current – "Corriente de resaca"
River mouth – "Boca del rio"	River current – "Corriente fluvial"
Rode – "Cabo o cable del ancla"	Roach – "Alunamiento"
Roll – "Balancearse"	Roger – "Recibido"
Roller furling – "Enrollador de vela"	Roller – "Onda grande, marejada"
Rooster tail – "Cola de gallo"	Roller reefing boom – "Botavara con enrollador"
RTV – "Sellador de silicón RTV"	Rough bar – "Barra con aguas agitadas"
Rudder – "Timón"	Rubrail (strake) – "Galón, verduguillo"
Running – "Navegar de empopada, correr"	Rudder stock – "Mecha del timon"
Running lights – "Luces de situacion"	Running fix – "Toma sucesiva de demoras de punto fijos"
Running rigging – "Jarcia de labor"	Running lights (vessels) – "Luces de posicion"
Sail area – "Superficie velica"	Safe water marks (fairways, mid-channels) – "Senales de aguas navegables"

Saloon (salon) – "Salon"	Sailboat – "Velero"
SAR emergency phases – "Fases de nivel de emergencia SAR"	Sampson post – "Bita Sampson"
SAR Mission official Coordinator (SMC) – "Responsible de coordinar la operación SAR (SMC)"	SAR incident form/folder – "Forma/carpeta del incidente SAR"
Schooner – "Goleta"	Satellite navigation – "Navegación por satélite"
Scouring – "Método de poner a flote una embarcacion encallada, utilizando la ehlice de la otra para abrir paso en el fondo"	Scope – "Alcance (de una cadena o cabo de ancla)"
Scupper – "Imbornal"	Screw – "Hélice"
Scuttle (to) – "Hundir intencionalmente"	Scuttle – "Portilla"
Sea chest – "Cajon de toma de mar; maleta de marinero"	Sea anchor – "Ancla flotante"
Sea cock – "Válvula (pasacascos) de toma de mar"	Sea chest – "Gate valve valvula del cajon de toma de mar"
Seabed – "Corriente marina"	Sea current – "Corriente marina"
Search pattern – "Pratron de busqueda"	Search and Rescue Unit (SRU) – "Unidad de Busqueda y Rescate (SRU)"
Seaworthy – "En buenas condiciones marineras"	Seaward – "Hacia la mar"

SECURITE – "SECURITE"	Secure (to) – "Hacer firme"
Self-draining cockpit – "Banera autodrenaje"	Seize (a line) – "Trincar (cabo)"
Sextant – "Sextante"	Set (of a current) – "Curso (de la corriente)"
Shaft – "Eje"	Shackle – "Grillete"
Sheepshank knot – "Nudo margarita"	Sheave – "Polea, roldana"
Sheet – "Escota"	Sheer – "Arrufadura"
Sheet track – "Guia de escota"	Sheet bend – "Vuelta de escota"
Shelter – "Abrigadero"	Sheet traveler – "Abrigadero"
Ship – "Buque"	Shift – "Palanca de cambio"
Shock load – "Incremento brusco de la resistencia en el cabo de remolque a causa del viento o del mar shore"	Shoal – "Bajío"
Shrouds – "Obenques"	Short range aids to navigation – "Ayudas a la navegacion de corto alcance"
Silence fini – "Silence fini"	Sidelights – "Luces laterales"

Sink – "Hundirse, naufragar"	Silence – "Silence"
Siren – "Sirena"	Sinker – "Peso muerto"
Skeg – "Talón de codaste"	Situation Report (SITREP) – "Reporte de situacion"
Slack wáter – "Marea de repunte"	Slack (to) – "Amollar"
Sling – "Eslinga"	Sliding Hatch – "Escotilla deslizante"
Slip clove hitch – "Ballestrinque escurridizo"	Slip – "Puesto de amarre en una marina"
Smoke and illumination signal – "Bengala combinada (humo y pirotecnica)"	Sloop – "Balandra"
Sole – "Suelo"	Snap shackle – "Mosquetón"
Sound signal – "Señal acústica"	Sound buoys – "Boyas con equipo acústico"
Spare part – "Repuesto"	Sounding – "Sondeo"
Spars – "Perchas"	Spark plug – "Bujía"
Spinnaker (symmetric/asymmetric) – "Espí (simétrico/asimétrico)"	Special purpose buoy – "Boya especial"
Splice – "Empalme, ayuste"	Spinnaker pole – "Tangon del espí(nnaker)"

Spring line – "Esprin"	Spreader – "Cruceta"
Squall – "Chubasco"	Spring tide – "Mares vivas"
Square knot (reef knot) – "Nudo llano"	Square daymarks – "Marcas diurnas cuadras"
Stanchion – "Candelero"	Square rigged – "Aparejo de vela cuadra"
Standard navy preserver (vest type with collar) – "Chaleco salvavidas estándar de la Marina (con cuello)"	Stand on vessel – "Embarcación que sigue a rumbo"
Standing rigging – "Jarcia firme"	Standing by on channel 16 – "Sigo en escucha canal 16"
Starboard hand – "Mark señal que se deja a mano derecha"	Starboard – "Estribor"
Static electricity – "Electricidad estática"	Starboard tack – "Amurado a estribor"
Station buoy – "Boya de posición"	Static forces – "Fuerzas estáticas"
Stay – "Estay"	Station keeping – "Mantener la embarcación en posición estacionaria"
Steerage – "Gobierno del timón"	Staysail – "Trinquetilla"
Steering Wheel – "Rueda del timón"	Steerageway – "Velocidad mínima de gobierno"

Stem pad eye (trailer eye bolt) – "Cáncamo en la roda (cáncamo de remolque)"	Stem – "Roda"
Stern – "Popa"	Stem the forces – "Compensar viento y corriente"
Stern navigation light – "Luz de alcance"	Stern line – "Largo de popa"
Storm jib – "Tormentín"	Stokes litter – "Camilla Stokes"
Stowage – "Estiba"	Stove (gimbaled) – "Estufa (con suspensión universal)"
Stranded rope – "Cabo de cordones"	Stowage locker – "Panol de estiba"
Strut – "Abrotante del eje"	Strobe light – "Lampara estroboscópica (luz intermitente de alta intensidad)"
Surf – "Rompiente"	Superstructure – "Superstructura"
Surf Rescue Boat (SRB) – "Embarcación de rescate (SRB)"	Surf line – "Linea de rompientes"
Surface swimmer – "Nanador de rescate (SAR)"	Surf zone – "Zona de rompientes"
Sweep Width (SAR) – "Medida de avistamiento"	Survival kit – "Kit de soberviviencia"
Swells or seas – "De la marejada (swells), del oleaje (seas)"	Swell – "Mar tendida"
Swimmer's harness – "Arnés de seguridad para nadadores de rescate"	Swim platform – "Plataforma de nado"

Switch to channel 68" – "Pase al canal 68"	Switch – "Interruptor"
Tack – "Bordada"	Tachometer – "Taquímetro"
Tackle – "Aparejo"	Tack (to) – "Virar, dar bordadas"
Taffrail – "Coronamiento de popa"	Tactical diameter – "Diámetro táctico"
Tank – "Tanque"	Tandem – "En fila india"
Thimble – "Guardacabo"	Telltale – "Catavientos"
Throttle – "Acelerador"	This is – "Aquí"
Through bolt – "Pasador"	Throttle lever – "palanca del acelerador"
Tidal current – "Corriente de marea"	Thumbs up – "Gesto del visto bueno (VoBo)"
Tighten (to) – "Cazar"	Tie down – "Asegurar con amarras"
Time zone – "Huso horario"	Tiller – "Cana del timón"
Toed (Toed in) – "Convergencia de la proa del remolque con la del barco remolcador"	Toe rail – "Regala"
Topping lift – "Amantillo"	Topmarks – "Marca del tope"

Tow line – "Cabo de remolque"	Topside – "Cubierta"
Towing hardware – "Herrajes de remolque"	Tow strap – "Esprín de remolque"
Track – "Trail"	Towing watch – "Vigía del remolque"
Traffic Separation Scheme – "Dispositivos de separación del tráfico"	Track Spacing – "Distancia entre rutas paralelas de búsqueda (SAR)"
Transducer – "Transductor"	Trail line (tag line) – "Cabo guía"
Transom – "Espejo de popa"	Transformer – "Transformador"
Triage – "Triage"	Trawler – "Buque arrastrero"
Trim – "Trimado"	Triangular daymark – "Marca diurna triangular"
Trimaran – "Trimarán"	Trim control – "Sistema de trimado"
Trough – "Seno de la ola"	Tripping line – "Espía"
True wind – "Viento real"	True course – "Rumbo verdadero"
Two-stroke engine – "Motor de dos tiempos"	Turnbuckle – "Tensor"
Under way – "En travesía"	U.S Aids to Navigation System – "Sistema de EEUU de Señalización Marítima"

Uniform State Waterway Marking System (USWMS) – Sistema estatal de señalización aguas del interior (USWMS)"	Underhung rudder – "Timón suspendido"
Utility Boat (UTB) – Bote de servicio de la Guardia Costera"	Universital Coordinated Time (UTC) – "Tiempo Universal Coordinado (TUC)"
Variation – "Declinación magnética"	Vang – "Osta"
Ventilation – "Ventilación"	Vari-nozzle – "Lanza de contraincendios"
Venturi effect – "Efecto Venturi"	Ventilator – "Ventilador"
VHF/HF Radio – "Radio VHF/HF"	Vessel – "Embarcación"
Wake – "Estela"	Waist and/or tag lines – "Amarras de costado para remolque"
Watch circle – "Campo de giro (de una embarcación fondeada o de una boya)"	Warm front – "Frente cálido"
Water gauge – "Indicador del nivel de agua"	Water capacity – "Capacidad de aqua dulce"
Water system – "Servicio de agua"	Water pump – "Bomba de agua"
Waterline length – "Eslora en la línea de flotacion"	Water tank – "Tanque de agua"
Wave – "Ola"	Watertight integrity – "Hacer estanco"

Wave height – "Altura de la ola"	Wave frequency – "Frecuencia de la olas"
Wave length – "Longitud de la ola"	Wave interference – "Interferencia de la olas"
Wave reflection – "Refección de las olas"	Wave period – "Periodo de la olas"
Wave saddle – "Seno de la ola"	Wave refraction – "Refracción de las olas"
Wave shoulder – "Hombro de la ola"	Wave series – "Serie de olas"
Weather Helm – "Cana a barlovento"	Weather (to) – "Hacia barlovento"
Well deck – "Cubierta de pozo"	Wedge – "Cuna"
Wet suit – "Traje húmedo"	Wet locker – "Panol de servicio"
Whistle – "Silbato; sirena"	Whipping – "Falcaceadura"
Williamson turn – "Virada de emergencia Williamson"	Whistle buoy – "Boya con sirena"
Winch handle – "Manivela del chigre"	Winch – "Chigre, guinche"
Wind instrument – "Anemómetro"	Wind direction – "Direccion del viento"
Wind vane – "Veleta / grimpola"	Wind shadow – "Al resguardo del viento"

Wind-chill factor – "Factor de enfriamiento del viento"	Wind's eye – "Filo del viento"
Window – "Ventana"	Windlass – "Molinete"
Yaw – "Dar guiñadas"	Windward – "A barlovento"
	Yawl – "Yola"

Vocabulario Aeronáutico/ Aeronautical Vocabulary

A catch in a tie rod keyway – "Un dispositivo de cierre en la ranura de la barra de acoplamiento (o biela)"	A sponsoring gas turbine owner/operator – "El auspicio de un propietario u operador de turbinas de gas"
A&P certificate – "Licencia A&P (fuselaje y motor)"	A sponsoring gas turbine owner/operator – "El auspicio de un propietario u operador de turbinas de gas"
Able Bodied Assistant – "Asistentes que esten en Buena condición física"	Abeam – "Por el través"
Accountable forms – "Formularios de rendición de cuentas"	Access to ramp – "Acceso a la rampa"
Actual cost – "Precio real"	ACMI/DAMP Agreement – "Ver explicacion"
Aero gas turbine engine – "Motores con turbinas de gas de aviación"	AD status – "Estado de las directivas de aeronavegabilidad"
Aerobridge – "Pasarela de acceso directo, pasarela de embarque/desembarque"	Aero gyro technology – "Tecnología de giro aéreo"
Aft facing – "Mirando hacia la popa (la parte trasera del avión)"	AFT – "Posterior"
Air craftsman – "Especialista en aeronáutica"	Aftermarket – "Mercado de accesorios"
Air return, gate return – "Retorno aereo, retorno a la puerta de salida"	Air gasper – "Dispositivos de control de aire"

Airborne radar – "Radar aerotransportado"	Air-washed – "En contacto con el aire"
Aircraft (plural) – "Aeronaves"	Aircart – "Generador externo/unidad externa suplidor/ra de aire"
Aircraft Clearance Zone – "Zona libre de obstáculos (para aeronaves)"	Aircraft Bridging – "Transición de aeronaves"
Aircraft Derivative Gas Turbines – "Turbinas de gas aeroderivadas"	Aircraft complaint cards – "Tarjetas/formularios de reclamaciones del avión/aeronave"
Aircraft gate utilization fee – "Cuota de derecho de uso de puerta de entrada"	Aircraft engine división – "División (gerencia/área) de motores para aeronaves (de aviación)"
Aircraft jacking – "Levantamiento, elevación"	Aircraft heaters – "Sistemas de calefaccion (para aeronaves)"
Aircraft revene time – "Tiempo productive (o tiempo de generación de ingresos) de la aeronave"	Aircraft On Ground (AOG) – "Aeronave en tierra"
Aircraft skins – "Revestimiento de la aeronave"	Aircraft sealing – "Cierre hermetico de aeronaves"
Aircraft strike – "Impacto que afecta a una aeronave"	Aircraft stand – "Plataforma de estaciónamiento"
Airfield how as and on routs of flights – "Tanto en las proximidades del aeropuerto como en las rutas de vuelo"	Aircraft Vessel Report – "Manifiesto/relación/listado de pasajeros"
Airplane single engine land – "Avion monomotor terrestre"	Airplane hanger – "Hangar"

Airport operators and handlers – "Operadores y administradores de aeropuertos"	Airport "in plane" vendor – "Proveedor de combustible del aeropuerto donde se encuentra la aeronave"
Airway – "Vías respiratorias"	Airport-pair – "Par de aeropuertos"
Albuterol – "Inhalador de albuterol"	Airworthiness – "Aeronavegabilidad"
Alu shot peened – "Aluminio granallado"	All out subject – "(Dentro de los 60 minutos de) la salida general, sujeto a las exigencias del trafico"
Amplitude video, is displayed with synthetic video – "Video de amplitude, se representa... con video sintético"	Amidship – "Al medio del buque o barco"
And auxiliary power unit – "Y la unidad de potencia auxiliar / y el grupo electrogeno de pista"	And accept the equipment by waiver – "Y aceptar el equipo mediante tal renuncia"
Angle of lead – " Ángulo de avance"	And tracking limited – "Limitada en su movimiento"
Anti-servo tab – "Aleta anti-servo/antiservo"	Antenna azimuth side lobes, sub-clutter – "Lobulos laterals de la antena de azimut (acimut), visibilidad de ecos perturbadores de radar"
Approach plate – "Carta de aproximación"	APP (Conditions off light) – "APP – Numero de Aproximaciones"
Apron Safety Handbook – "Manual de seguridad en plataforma"	Apron – "Plataforma/rampa de parqueo/estacionamiento"

Armature and slator wiring – "Cableado del inducido y el estator"	Arc-track resistant wires – "Cables resistentes a arcos electricos (de alto voltaje)"
As of the more – "Desde / la que fuera mas reciente"	As expiring – "A su vencimiento"
Astrogator – "Navegante de una nave espacial"	Asset container – "Contenedor aéreo"
At a bank – "En ángulo / inclinado"	Astronautical navigation star display – "Representación estelar para la navegación astronáutica"
Automatic choice of mínimum jammed frequency – "Selección automatic de frecuencias con menor interferencia"	Authority: a power that a person is vested with – "Autoridad: Poder que se otorga a una persona"
Automatic pitch trim – "Compensación automática de pitch"	Automatic External Defibrilator (AED) – "Desfibrilador externo automático (DEA)"
Aviation scout – "Exploradora a , reo"	Avery Hardoll Inlet Coupler – "Cople o acoplamiento de admisión"
Avionics bay equipment rack – "Bastidor de equipo en el compartimento principal de aviónica"	Avionics – "Aviónica"
Avoidance – "Evitacion, medidas evasivas (ver enlace a la Ley)"	Avionics ramp – "Control de instrumentos en rampa"
Award for AVIA D Poznan Fear – "Premio para AVIA D en la Feria de Poznan"	Avulsion – "Avulsión / denudación [traumática] / colgajo"
Single reconnaisance squadrons – "Escuadrones de reconocimiento aéreo"	Axle jack – "Gato/soporte regulable en altura del eje"

Backcast – "Retrospectivo"	Back plate – "Placa posterior o trasera"
Baggage On Hand (BOH) – "Equipaje de mano"	Backshell – "Escudo trasero"
Balancing of aircraft gauge – "Medidor de balance de la aeronave"	Balance – "La balanza"
Bank strucutre (aviation) – "Estructura del banco (de vuelos)"	Ball valve strainer with glass window and reverse feature for defue – "Filtro de la válvula de bola con mirilla y opción inversa para purga"
Battleship – "Conete de pruebas"	Baseline requirements – "Requisitos básicos / de base / mínimos / fundamentales / etc."
Bearing – "Rumbo"	Beacon only target reports – "Informes de objetivos exclusivos de balizas"
Bellcrank (wings) – "Palanca acodada"	Beaver Creel Trout fabric – "Tela suave tipo piel de castor"
Bench test per CMM (company maintenance manual) – "Pruebas de banco por CMM (manual empresa mantener)"	Belly belt – "Cinturón abdominal"
Big Daddy – "Monstruo"	Bending brake – "(Prensa) Plegadora"
Bladder layer – "Capa de kevlar"	Blackout date – "Fechas restingidas (periodos de cierre / fechas de prohibición / periodos de interrupción)
Blanket third-country code share authority – "Autorizacion general para comparti códigos con terceros países"	Blades track/lead-tag – "Paso/avance-retraso de las palas"

Blockspace – "Bloqueo de espacio"	Block in/block out – "Tiempo de parqueo"
Bobtail – "Rabón"	Blow molding – "Moldeo o moldeado por inyección de aire / moldeo por soplado"
Bond – "Ajustar / instalar"	Body-bound bolted joint – "Empalme (o junta) empernado (unido mediate pernos) al cuerpo"
Bonded doublers – "Chapas de refuerzo soldadas por adherencia"	Bond supervisor – "Supervisor de existencias"
Bone shaft – "Diafisis"	Bonding – "Metalización"
Booster – "Propulsor / (cohete) impulsor"	Boom hanger – "Soporte del brazo extensible"
Borescope – "Boroscopio"	Booster style seats – "Asientos tipo 'booster'"
Boss fitting – "Empalme de resalte"	Boresighted weapon – "Arma visada por el animo"
Box section bladder tank – "Tanque tipo vegija de la sección de la caja del plano/ala"	Buoyancy – "Flotabilidad"
Break rider – "Operador de maniobras"	Brace – "Preparse para el impacto"
Briefings and debriefings – "Briefing y debriefings"	Breech-loaded gas turbine – "Turbina de gas con retrocarga"
Budget travel – "Vuelos/viajes de bajo costo/de costo reducido"	Bubble door – "Puerta de burbuja"
Bulk motion – "Movimiento general o principal"	Buffer – "Rango de tolerancia"

Bundle drop – "Lanzamiento de paquetes"	Bullet – "Carenado/carenaje en forma de proyectil"
Bus bar braids – "Enrejado conductivo"	Burn testing – "Prueba de combustión
Bute Doors and DFDR (Digital Flight Data Recorder) Repair – "Puertas Bute y reparación de recuerda digital data avión"	But for schedules – "Los programas 'but-for'"
c/n – "Numero de catalogo"	C-scan imaging – "Toma de imágenes c-scan"
Calling back – "Volver a llamar"	Cache – "Almacenan en la cache"
Canard-wing – "Ala canard"	Canard – "Aletas canard"
Capabilities – "Capacidades"	Cap strip – "Tiras de refuerzo"
Cardiopulmonary Resuscitation (CPR) – "reanimación cardiopulmonar"	Capacity Staggering – "Escalonamiento en capacidad"
CAVOK – "Nubes y visibilidad correctas"	Cargo Satellite Fueling Facility – "La instalación de avituallamiento de combustible"
Certificate Holding District Office – "Direccion regional"	CDI (Course Deviation Indicator) scaling – "Indicador de Desviación de Curso/Rubo (CDI)"
Chariman, council committee for joint support of air navigation services – "Comité de del Consejo de ayuda colectiva para los servicios de navegación aérea"	Cessna ProParts programme – "El programa ProParts de Cessna"

Check airman – "Instructor/piloto instructor/instructor de vuelo"	Chartered – "Contratado"
Child restraint systems – "Sistemas de sujeción infantil/para niños"	Chemical burns, termal burns, electrical burns – "Quemaduras de origen químico, térmico y eléctrico"
Circular approaches – "Aproximaciones circulares"	Chin bubble – "Burbuja/burbuja inferior/venta burbuja de los pedales/ventana burbuja"
Civilian Chief – "Jefe civil"	City pair – "Ciudades de salida y de destino"
Clear of any ingestión or engine exhaust área – "Lejos de cualquier zona de ingestión o escape del motor"	Clean stall speed – "Velocidad neta/franca de perdida de sustentación "
Clecoes –"Clecos"	Clearance lights with guards – "Luces de galibo con cubiertas de seguridad"
Close Quarters Countermeasures – "Sistemas de combate cuerpo a cuerpo"	Clock position – "(Posiciones) como en el reloj"
Close tolerance – "Con tolerancia mínima (o estricta)"	Close suction – "Cierre/a la aspiración"
Close-out inspection – "Inspeccion final (de cierre)"	Close-air-support – "Apoyo aéreo cercano"
Closest point of approach (CPA) – "Punto mas cercano de aproximación"	Close-out to the Load Planner – "Cierre de vuelo para el despachante de la aeronave"
Coast – "Orbita"	Closing time / opening time orifice – "Orificio (del control) de tiempo de apertura y cierre"
Coat rod – "Barra para colgar perchas / Barra para colgar ropa"	Coasted track – "Pista proyectada/extrapolada"

Cold gaz Ejector Release Unit – "Unidad de lanzamiento/liberación/descarga del eyector"	Cockpit mock-up – "Maqueta de cabina de mando"
Collar stop – "Obturador de collarín"	Cold weather start – "Sistema de arranque en frio"
Common carriage – "Servicio publico de transporte"	Commercial-Off-The-Shelf Product – "Productos disponibles comercialmente / productos del desarrollo"
Compact Disc Drive Unit – "Unidad de discos compactos/lectoras de discos compactos"	Commuter aviation industry – "Industria de aeronaves de tamaño mediano (o pequeño)"
Completion center – "Planta de acabado"	Comped – "El sector de compensaciones"
Contact stand – "Gate/puentes de embarque"	Consolidation , deconsolidation – "Consolidación, deconsolidacion (en servicios de carga)"
Contiguous United States – "Los Estados Unidos contiguos/Estados Unidos continentales (sin Alaska)/Los 48 de abajo"	Container Transfer Loading Pre-arrival – "Carga/transporte para transferencia del container antes del arribo [del avión]"
Contract advertised and one vendor participated – "Contrato fue anunciado/publicado y solo un vendedor presento una oferta"	Continuous wrought material – "Forjado en una sola pieza"
Coordinate accuracy and discrimination – "Precision de coordenadas – discriminación"	Contract Department – "Departamento de contratos"
Corporate aircraft – "Aeronaves corporativas"	Copper flashing over the metal backing – "Tapajuntas de cobre sobre el sporte metálico"

Coupling balance journal ground – "Acoplamiento del cojinete de balanceo en alineación concéntrica"	Country sourcing – "Aumenta las compras en países con costos competitivos"
Course and clearance transmitters – "Transmisores de curso y clearance / de cobertura lateral"	Coupon lifts – "Cuponse levantados"
Creep análisis – "Deformación bajo carga constante a lo largo del tiempo / deformación continua bajo carga constante"	Crash fire fighting – "Equipos contra incendios en caso de accidentes / equipos de extinción de incendios de accidentes"
Critical aircraft situation training – "Entrenamiento para situaciones criticas en la aeronave"	CRES (Corrosion-Resistant Steel) – "Acero resistente a la corrosión"
Crows-feet – "Pata(s) de gallo"	Crown (of a ship) – "Corona/Corona de barbotin"
Cruise torque – "Torsión de vuelo"	Cathode ray tube (CRT) – "Tubo de rayos catódicos"
Cueing & Slaving – "Localización y captura"	Cubesat – "CubeSat"
Curriculum footprint – "Organización del curriculum"	Currency – "Actualidad"
Customs ullage cage – "Espacion/sección/contenedor vacio de la aduana; espacion no ocupado o disponible en la aduana"	Customs productions – "Artículos hechos de encargo privado/producidos a la especificación de un cliente"
Cyanosis – "Cianonsis"	Cut-off and availability times – "Plazos de entrega y disponibilidad"

Dampener plate – "Placa amortiguadora"	DA performance – "rendimiento del controlador de descenso"
Deadhead – "Que no están en funciones"	Dash speed – "Velocidad máxima"
Deplaning – "Bajarse del avión, apearse"	Deep space – "Espacio lejano/profundo"
Designation number – "Numero de designación"	Derived Flow cycle – "Ciclo de flujo derivado"
Desktop and airport services – "Servicios aeroportuarios y de atención al cliente"	Desing day (en este contexto) – "Dia de diseño"
DFO (Director Flight Operations) – "Director de Operaciones de Vuelo"	Detected target range – "Determinar la distancia al objectivo"
Digital signal processing computer calibration routine – "Rutina de calibración computarizada para el procesamiento de señales digitales"	Digital Mission Engineering (DME) – "Ingeniería digital de misión"
Dimpling tool – "Herramienta de avellanado"	DIM Tracer – "Bala trazadora visible solo con visores nocturnos o de infrarrojos / trazadora nocturna"
Dirty box – "Caja de protección / aislante"	Diphenhydramine – "Difenhidramina"
Dirtying up the wing – "Ensuciar el avión / pasar el avión a configuración sucia"	Dirty finger print certification – "Certificación/aprobación del récord de mantenimiento (aviones)"
Disruption – "Interrupción / altercaion / perturbación"	Display – "Monitor"

Divan in Aft – "Diván en la parte trasera de la cabina (de pasajeros)"	Distress tracking – "Seguimiento de aeronaves en peligro / emergencia"
Dog fight – "Combate aéreo cercano"	Do not reduce power until in the flare – "No reduzca la potencia hasta que rompa el planeo / hasta la nivelada"
Door steward – "Portero"	Doghouse – "Compartimento auxiliar"
Downline and layover – "Bajada y parada"	Double frequency diversity – "Doble diversidad de frecuencia"
Downline catering – "Servicios de cocina de vuelo"	Downline bumping – "Cambiar / le / cancelar / le el cuelo (a un viajero)"
Downward drag – "Tracción / resistencia aerodinámica descendente"	Download training from more expensive aging training aircraft – "Transferir el entrenamiento desde aviones mas viejos y caros"
Drain mast – "Mástil de drenaje"	Drag link spring – "Resorte de unión de arrastre"
Drilled wing fitting – "Herraje de sujeción de las alas perforado / taladrado"	Dress covers – "Forros de los asientos"
Driving – "Mando / maquina impulsora"	Drive away – "Colisiones que no impiden la marcha"
Dropping low – "Descender hasta altitud baja"	Drivingly – "(Acoplado) de modo / manera accionante (a un generador eléctrico)"
Dry stores – "Almacenes (o contenedores) de alimentos secos"	Dry leasing (aviation) – "Arrendamiento / alquiler sin tripulación"
Dual boom microphone headsets – "Juego de auriculares con micrófono de doble	DU (acronym) – "Documento Único"

reverberación"	
Dump cooled exit cone – "Cono de salida con enfriamiento por descarga"	Dual jack screw type actuators – "Actuadores dobles tip tornillo sin fin"
Dynamic geofencing – "Geoperimetraje dinámico"	Dust flinger – "Elimador del polvo"
E-plars / singar systems – "EPLARS con sistema de arrastre"	E- path lighting – "Iluminacion de evacuación / de emergencia"
Earth short / contact short to earth – "Cortocircuito a tierra / corticircuito por contacto a tierra"	EA-18G ACAT 1D Milestone B Decision Review – "Revision decisoria de la etapa B del ea-18g acat 1d"
EGT margin – "Margen de EGT (temperatura de los gases de escape)"	Eastern/Western Range – "Zona de lanzamiento de la Costa Este/Oeste"
EkW – "Kilovatio equivalente"	Ehp – "Caballos de Vapor efectivos"
Electrolitic cell – "Celda electrolítica"	Elbow from Avery – "codo (marca) Avery"
Emergency ditching – "Amarizaje forzoso / de emergencia"	Electronic flight strip – "Sistema electrónico fichas de progreso de vuelo"
End-on end combinations – "Combinaciones triangulares"	Encumbrance – "Afectación"
Engine nacelle área – "El área de la barquilla o góndola de motor"	Engine exhaust cover – "Tapa del escape del motor"
Engine reserve – "Combustible"	Engine rebuilt "0" time – "Motor reconstruido en fabrica y llevado a cero"

EOS-era validation – "Validacion en el marco (en la era) del Sistema de Observación Terrestre [EOS]"	Engine torching – "Fuego a la salida del motor"
Ethernet capable MFD – "MFD habilitado para Ethernet"	Equipment fit – "Equipamiento"
Exhaust Valve Gap / front running light / steering / dipstick – "Boca de la válvula de escape"	Exceedencies – "Excesos"
Expansive mission radio capabilities – "Capacidad de comunicación en una gran cantidad de frecuencias"	Expansion type or ground down solid reamer – "Escariador extensible o fijo de banco"
Experimental and Kit planes – "Aviones de kit y experimentales"	Expectations – "Expectativas"
Extended holding on runway – "Tiempo de espera adicional en pista"	Expertise (in this context) – "Pericia / técnica / destreza /maestría"
Eye gantry – "Pórtico de Ojo / Pórtico de Observación"	Extended overwater turbojet-powered airplane operations – "Operaciones extendidas sobre el agua de avión turbojet/ / turborreactor"
Fan blades – "Alabes de fan"	Eye-rings –"Ojos de cable"
Feathering spring to change the pitch – "Y al resorte de embanderamiento cambiar el paso de la hélice"	Fare carrying flight – "Vuelo de transporte de pasajeros"
Feed to – "Al lado a alimentar"	Feed – "Alimentar / vuelos de alimentación / flujo / volumen (depasajeros)"

Fielded – "Desplegados"	Ferry flight – "Vuelo ferry"
Flame-tube – "Tubo de llama"	Flag operations – "Operaciones / vuelos internacionales"
Flap piano hinge – "Bisagra tipo piano del flap"	Flap bay – "Compartimiento de flap"
Flight attendant – "Auxiliar de vuelo"	Flat rated – "Empuje nominal"
Flight deck – "Cabina de vuelo"	Flight crossing – "Vuelo de travesía"
Flight into known icing – "(Certificado para) volar en condiciones de hielo / engelamiento conocido"	Flight departure and closeout – "Salida del vuelo y cierre"
Flight reléase – "Hoja / información de despacho del vuelo"	Flight progress strips – "Fichas de progresión de vuelo"
Float plane – "Hidroavión"	Flight share – "Vuelo de costo compartido"
Fluorescent penetrant – "Inspección mediante penetrante fluorescente"	Flow-to-open, Flow-to-close – "Flujo de cierre , flujo de abertura"
Fly away – "Anulación del registro de vuelos de salida / despegue"	Flush repair – "Reparación al ras de la superficie"
FMGEC / Flight Management Guidance and Envelope Computer – "FMGEC –	Flyaway equipment – "Equipo listo para el vuelo (avión de fabrica)"

Computadora de Guiado, Envolvente, y Gestion de Vuelo"	
Foot – "Pie (de montaje)"	FODO – "Oficial de turno de operaciones aéreas"
For proper climb and heading instructions – "Para tener las instrucciones adecuadas / apropiadas de ascenso y rumbo"	For compensation or hire – "Por remuneración o contrato"
Forward breakaway power – "Zona de peligro al romper inercia de rodaje"	Forklift pockets – "Cavidades / huecos / alojamientos para montacargas / carretillas elevadors"
Forward Spar Fitting – Machined and Drilled – "Herraje del larguero delantero – Maquinado y perforado"	Forward fit – "Equipados en aeronaves nuevas"
Forward-looking information is subject to risk and uncertainty – "Información prospectiva sujeta a riesgo e incertidumbre"	Forward unit – "Unidad de avanzada"
Free air temperatura (plane) – "Indicador de temperatura del aire (dentro del avión)"	Four-place pistón – "De cuatro plazas, con motor a pistones"
Freewheel – "Rueda libre"	Freeflow – "Vuelo libre"
Fuel agent – "Abastecedor de combustible"	Fresh hot section – "Sección caliente nueva"
Fuel fluid management – "sistema de administración de combustible / sistema de control de combustible (según el país)"	Fuel drop tank – "Desposito de combustible desechable"

Fuel sense line air bleed valve – "Válvula de purga de aire de la línea del sensor de combustible"	Fuel operator – "Técnico / operario encargado del combustible"
Fuelling pit – "Zona / pit / fosa para el abastecimiento de combustible"	Fuel ticket – "Recibo / comprobante del combustible"
Functional dual controls – "Controles funcionales dobles"	Full-fare carriers – "Líneas aéreas convencionales"
Gas generator failint to achieve starter off speed – "El generador de gas no alcanza la velocidad de desconexión del arrancador"	Gammon short – "Gammon Corto"
Gear down landing – "Tren de aterrizaje desplegado"	Gas wash exhaust – "Conducto de salida de gases"
Give them attention, open ranks, okay? – "Ordene / De la orden de que tomen posición de revista, entiendo?"	Gimble – "Cana del timón"
Glareshield – "Antireflejo"	Gland housing – "Porta guarnición"
Go off – "Despegarse"	Go – "Preparado / listo"
Green configuration – "Sin pintura ni interior"	Godspeed! – "Buen viaje! Buena suerte!"
Gross terminal sizing – "Tamaño estimado de la terminal"	Green taxiing – "Rodaje ecológico"
Ground source – "Muestra extraída"	Ground and ingestión – "Modem terrestre / en tierra"

Ground-to-air radios – "Equipos de radiocomunicación tierra-aire / radios para comunicación tierra-aire"	Ground track – "Trayectoria sobre el terreno"
G's are starting to build – "Fuerzas G en aumento"	Growth potential – "Potencial de crecimiento"
Hammerhead – "Maniobra acrobática de escape"	Half time - "Medio ciclo (de recondicionar)"
Handpiece – "Extension portaherramienta manual"	Hamp packing – "Empaque sellado"
Hard time – "Tiempo critico"	Hangarkeeper – "Propietarios de los hangares"
Hardware heritage – "Inventario de hardware"	Hard time limited component – "Componente con limitación estricta de tiempo"
Head-up – "Líder / liderar / dirigir"	Haverfunctions – "Funciones Haver"
Headstock / tail stock – "Cabezal / contrapunto"	Headliner – "Tapizado del techo"
Heating, refrigerating – "Calefacción, refrigeración"	Hearing docket – "Registro de procesos, lista de audiencia, lista de litigios"
Heavy shop – "Taller"	Heavy check – "Verificación exhaustiva"
Heliguard – "Oficial de aterrizaje de helicópteros"	Hedge fuel – "Cobertura de combustible"
High contour – "De alto contorno"	Hi-lites – "Perno / remache tipo Hi-Lite"

High solids epoxy sanding surfacer – "Nivelador para lijar / pulir con (resina / esmalte / pintura) Epoxi de alto contenido en solidos"	High lift wing system – "Sistema de ala de alta sustentación"
Hold & bulk control – "Control de custodia & bultos / seguridad de equipaje y cargo"	High Surface part – "Alto nivel de acabado superficial"
Home star – "Sol"	Holds – "Esperas"
Honored – "Aceptados"	Homing radio navigational aids – "Vuelo hacia el objetivo por medio de instrumentos de navegación"
Horizon tail span – "Envergadura del timón de profundidad"	Hook-up fee – "Tarifa conexión"
Hot-and-high – "De gran activad y alto rendimiento"	Horizontal / vertical gradiometer booms – "Plumas de gradiometro vertical / horizontal"
Hovering out (of ground effect) – "Vuelo estacionario fuera del efecto suelo"	Housing – "Almacenaje o alojamiento"
Hub – "Centro (figurado) / (perno para) buje"	HSG Assembly – "Ensamblaje del generador de alta velocidad"
Hub installation straight bore low momento configuration – "Configuración para el momento bajo con perforación recta para la instalación del cubo"	Hub bore – "Orificio del cubo"
Hub-and-spoke system – "Estructura o sistema con centro de distribución de vuelos / sistema de centro y radios"	Hub pullup – "Distancia / nivelación del cubo con la barra"

Hypoxia – "Hipoxia"	Hub-to-hub routes – "Rutas de aeropuerto central a aeropuerto central"
Ident filters – "Filtros de reconocimiento / verificación de identidad"	ICD – "Controles y diagnósticos inteligente"
Impeller (aviation) – "Impulsor / propulsor / motor / rotor"	Idle thrust / idle reverse – "Empuje de ralentí / ralentí de reversa"
Inboard – "El interior, asientos en el área interior"	In-flight shut-down (IFSD) – "Apagado en vuelo del motor"
Inbound leisure markets – "Turismo receptivo"	Inbound and outbound legs of the hold – "Tramos / piernas de acercamiento y alejamiento del circuito de espera"
Ingestion – "Ingestión"	Incoming threat – "Amenaza en aproximación"
Instructional power cruise speed – "Velocidad de crucero con ajuste de potencia especial para la instrucción"	Installation of navigation sights in accordance with Master Drawing List – "Instalación de vistas de navegación de acuerdo al listado maestro de dibujos"
Instrument panel dimmer bus – "Barra atenuadora / de atenuación del panel de instrumento"	Instrument approaches (flight) – "Aproximaciones por instrumentos (vuelo)"
International Space Training Center – "Centro Internacional de Entrenamiento Espacial"	Insurance letter – "Comprobante para el seguro"
Integrated payload stack – "Paquete de carga útil integrado"	Integrated Airman Certification and/or Rating Application (IACRA) – "Certificacion integrada de aviadore y/o solicitud de valoración"

Ip check valve – "Válvula de detección de presión interna"	Internal body fuel tanks – "Depositos de combustible integrados en el fuselaje"
Joint Tactical Information Data Systems (JTIDS) – "Sistemas Data del Información Único Táctica"	Is rated and endorsed – "Esta calificado y firma"
Jet Carrier – "Compañía aérea"	Jet blast resistant – "Resistente al chorro (de gases) de los motores a reacción"
Jet way – "Manga de abordaje"	Jet engine – "Motor de reacción"
Jigging – "Plantilla de perforación / plantilla de cortar"	Jig holes – "Los orficios de la plantilla (útil) para/de las costillas (del avión)"
Jumpseat – "Asiento rebatible"	Joint and Edge covering – "Revestimiento de uniones y cantos"
K-loader – "K-loader"	Justification for Sole Source (Simplified Acquisitions<$100k) – "Justificación para único proveedor (adquisiciones simplificadas <$mil)"
Key principle – "Dispositivo con llava"	Kennel – "Cajas para transporte de animales"
KML Vector Tile Set and Volume Shapefile export options – "Opciones para exportar capas vectoriales en formato KML"	Kick-Start load – "Carga de arranque"
Kt – "Nudo"	Knot performance – "Capaz de desplazarse a una velocidad de nudos"
Land – "Parte plana /superficie"	LAC (Local Area Code) – "Código de área local"

Landing gear leg – "Pata del tren de aterizaje"	Land navigation facility performance – "Sistema de navegación"
Lap children – "Niños que viajan en el regazo / niños menores de"	Landing gear overhaul – "Revisión general del tren de aterrizaje"
Last-stage buckets – "Alabes en la ultima etapa, alabes de etapa final"	Last PC – "Ultima verificación de competencia"
Lb HP – "Libras por caballo de fuerza"	Lb st – "Libras estándar"
Leverage, 10 (in this context) – "Apalancamiento"	Left (and right) main – "Tren izquierdo (y derecho)"
LHS (lef-hand side) – "Lado izquierdo"	LH/RH book navigation chart cases – "Anaqueles para las cartas de navegación situadas a izquierda y derecha"
License certification subdirection – "Subdirección de certificación de licencia"	License up-rating flight – "Exámenes de vuelo para acceder a una categoría de licencia superior"
Lien – "Gravamen"	Liens, claims, and encumbrances – "Gravámenes, reclamaciones y afectaciones"
Line and base airframe – "Servicioes de mantenimiento de (la estructura del) fuselaje en/de línea y base"	Line position – "Funciones de línea"
Line stations – "Estaciones de línea"	Line-to-line – "Muy precisa (la tolerancia puede ser desde muy precisa hasta de varias milésimas de deviación)"
Line-up allowance – "Tolerancia de alineamiento/alineación"	Liveried – "Con los colores distintivos de"

Local Station Manager – "Jefe de estación local"	Lockable swivel – "Pivote inmovilizable"
Loiter – "(Cicuitos de) Espera"	Long-haul Budget services – "Servicios económicos/de bajo costo de larga distancia son vuelos regulares"
Long-line seismic support – "Apoyo en sismos utilizando línea larga"	Loose equipment – "Equipo suelto, desmontable, no fijo"
Loose oil ring – "Anillo suelto de engrase"	Lost Time Incidents – "Incidentes que ocasionan retrasos de tiempo"
Low-Signature Target – "Blancos de difícil detección"	M.D. / D.O. / P.A. / L.P.N. / L.V.N. / E.M.T. – "Doctor/Osteópata/Medico Especialista/enfermeras practica autorizada/auxiliar de enfermería/EME"
MAC backwards – "Mueve xxx hacia atrás la cuerda aerodinámica media"	Main deck loader – "Cargador / paletizador de / para la bodega principal"
Main landing gear bearing lugs – "Pernos de carga en el tren de aterrizaje"	Mainline airline – "Aerolíneas/compañías de aviación tradicionales"
Mainlines revenues Passenger – "La ganancia por pasajero de asiento por milla"	Maintenance Reserve Payments – "Pagos al fondo de reserva de mantenimiento"
Make and model – "Constructor y modelo"	Manned Space Rendezvous – "Acomplamiento Espacial Tripulado"
Marshalling –"Guía a base de comunicación visual/señalero"	Maryland Centre Tower requesting takeoff clearance – "La torre de Maryland Centre pidiendo autorización de despegue"
Material build-up – "Acumulación de material"	Maximum power assurance run – "Corrida a máxima potencia"

Medium batch – "Producción a mediana escala"	Meeters and mediana escala"
Meeters and greeters área – "Area/zona/sala de recepción para familiares de los pasajeros"	Meeting the Performance Group requirement – "Satisface el requisito de desempeño en grupo / en conjunto"
Mission avionic upgrade of military aircrafts – "Modernización de aviónica de misión en aeronaves militares"	Mobile Workers in civil aviation – "Personal de Vuelo (en aviación civil)"
Mod kit – "Kit de modificación"	Mouth-to-barrier – "Respiración boca a boca/a través del aparato respiratorio"
MRO maintenance repair and overhaul – "Mantenimiento, reparación y reacondicionamiento"	MTBF (Mean Time Between Failures / MTTF (Mean Time To Repair) data – "Datos tiempo medio entre fallos / tiempo medio para reparar datos"
Multi-carrier travel agencies – "Multi-proveedor, multiproveedor"	Multi-engine aircraft – "Aeronave multimotor"
Multi-Engine-Engine Pistón – "Aviones multimotores a pistón"	Multi-role/swing-role operations – "Operaciones con aeronaves multirole/ multimisión, y aeronaves que cumplen misiones simultaneas"
Multiple-burst – "Pulsos múltiples"	Nadir – "Nadir"
National Aeronautical and Space Administration (NASA) – "Administración Nacional de Aeronáutica y del Espacio"	Naval Air Training and Operations Procedures Standardization jackets – "Chaquetas de estandarización de procedimientos de operaciones y entrenamiento aeronaval"

NCh (No change) – "No hay cambios en el escaneo temico (Nch)/ No se han detectado nuevas fuentes de calor"	NDI (Nondestructive inspection) – "inspección no destructiva"
Near-space operations – "Operaciones en el espacio cercano"	Neck of the balloon – "Cuello del globo"
Negative G dives – "Picados a (para) gravedad negativa"	Net fly away Price – "Precio neto unitario 'fly away'"
Network planners – "Programadores / analista de rutas"	Network, Fleet, Guest Services, and Brand – "La Red (de aerolíneas)"
Neutralizes its momentum – "Neutraliza su propia inercia"	New comercial aircraft platform hydraulic engine-driven pumps – "Bombas hidráulicas accionadas por motor para la nueva plataforma de aviación comercial"
Night stop – "Parada/escala nocturna"	Nite seal; daily seal – "Sello nocturno, sello diario /diurno o cierre nocturno cierre diario/diurno"
Nitrated – "Nitrurado"	No more tan three threads extend beyond the turnbuckle – "Me parece que estas en lo correcto"
Non instrument approach surfaces – "Plano de aproximación visual"	Non-belted externally serviceable toilet – "Inodoro capaz de servicio exterior y sin cinturón de seguridad"
Non-flights – "No implicado (relacionado) con un vuelo"	Non-Fokker parts – "Piezas no originales de Fokker"
Non-return flap – "Válvulas (obturadores) antirretorno"	Non-rev crewmember – "Miembro de la tripulación que no ha pagado su pasaje/billete"

Non-roster day – "Días libres/días no asignados a vuelos"	Nonequilibrium ionization – "Ionizacion fuera de equilibrio"
Norwegian adapted Hawk / GDS 5 ton – "El sistema Hawk adaptado por Noruega / GDS de 5 toneladas"	Nose gear steering system – "Timón de tren de nariz"
NTP (Network Time Protocol) master clocks – "Reloje maestros que utilizan el protocolo de red de tiempo (NTP) para su sincronización"	Nose-up – "Subida del morro"
Obstacle climb – "Superación de obstáculos"	Nut, clip-on – "Tuerca con pinza/ tipo prendedor"
Of the ultimate load – "De la carga máxima"	Of the rover – "De la flecha"
Offline airport – "Aeropuerto fuera de línea"	Off – "De / desde / partiendo de / saliendo de / parte de / sale de"
Oil rubber – "Dispositivo de lubricación realizado en caucho (o goma)"	Offshore helicopter – "Helicóptero para operaciones costa afuera"
On its charter and scheduled network – "En su red de vuelos chárter y regulares"	Oleos – "Patas del tren de aterrizaje"
Open jaw – "Circuito abierto"	Onboard storage batteries – "Beteria de almacenamiento a bordo"
Operational Parachute Wings – "Alas de Paracaidista (en activo)"	Open-shies – "Cielos abiertos"
Order assessing civil penalty – "Resolución que impone una multa administrativa"	Operations specifications – "Especificaciones de operación"
OSM (Operations Safety Manager) – "Gerenta de Seguridad de Operaciones"	Originating leg – "El primer tramo/la primera escala del vuelo"

Out of autoclave processes – "Procesos posteriores al curado en autoclave"	OTQ (On Target Quality) – "Calidad objetivo"
Out of the rest – "Fuera de su base"	Out of limits fuel and hydraulic and oil leaks – "Escapes de combustible y fluido hidráulico por fuera de limites"
Outboard and tracked full aircraft forward – "Hacia el exterior, bien alineados y recorridos totalmente en dirección de la parte delantera del avión"	Out-station – "Fuera de la estación propia"
Outstation – "Estacion externa"	Outside stem and yoke – "Válvula de compuerta externa"
Overhead passenger service units (PSU) – "Unidades de servicio al pasajero"	Over wing – "Situada sobre el ala"
Overtorque and torque – "Sobretorque y torque"	Overhung weight – "Momento flector"
Pairing – "Itenerario"	Owner produced parts – "Partes fabricadas por el propietario"
Pallet – "Pallet"	Pairings – "Acope (de vuelos)"
Parafoil – "Parapista"	Pampas – "Producto de las pampas"
Passenger address tape reproducer – "P.A. con mensajes pre grabados"	Parallel divestment – "Presentacion paralela de equipaje de mano"
PC-base ATC Controller Workstation – "Estación de Trabajo del Controlador de Transito Aéreo basado en PC"	Payload – "Carga útil"

Percentage in weight – "Porcentaje en paso"	Peak hour thoughput loads – "Carga de trabajo en horas pico que afecta al equipo, procedimientos y otras áreas de seguridad"
Person directly affected – "El afectado directo"	Performance-secondary flow-meanline análisis – "Análisis de flujo de línea media de rendimiento secundario"
Phone "patch" – "Puente/relevo telefónico"	Personnel screening technologies – "Tecnologías de control de seguridad del personal / de personas"
Pick power, MTI (Moving Target Indicator) – "Potencia de pico (transmisión) , MTI o Indicador de blancos móviles"	Phrase on a safety card for a plane – "Face seat aft / track"
Pickup lens – "Lente lectora"	Pick-up wire – "Cable de captación"
Pilot action (forces on the controls) – "Acción del Piloto (Fuerzas aplicadas a los controles)"	Pintle-mounted – "Mondado(s) sobre pivote(s)"
Piston general aviation aircraft that use 100 LL fuel – "Limitar las emisiones de plomo de los aviones a pistón de uso general que utilicen combustible 100LL"	Piston-driven – "Impulsado por motor a pistón"
Pitch – "Espacio entre los asientos"	Pitch-links, tabs – "Conexiones del ángulo de ataque, estabilizadores"
Placarded – "Etiquetados"	Planetary gears – "Engranajes planetarios"
Plug in gauge for nozzle – "Manómetro de clavija para boquilla"	Plug together wiring loom – "Mazo de cables totalmente enchufable/con conectores

	enchufables"
Pneudraulics – "Sistemas hibridas neumática-hidráulica"	Pneumatic de-icing boots – "botas neumáticas para deshielo"
Pod-and- boom configuration – "Configuración de góndola / compartimento y larguero"	Pool engine – "Reserva de motores de repuesto"
Poppet stem – "Vástago de resorte"	Positioning – "Tiempo de psocionamiento"
Positioning/depositioning – "Emplazamiento/retiro (Aviones, barcos, y naves espaciales)"	Positive space business pass – "Pase con lugar confirmado en clase ejuctiva"
Postbooster – "Post-aceleradores/post-impulsores"	Pot-bellied pigs – "Cerditos enanos/diminutos (vietnamitas)"
Pre-flight/post-flight editing – "Revisión previa y posterior al vuelo"	Prepreg – "Prepreg, preimpregnado"
Pressure cowls – "Capo presurizado"	Pressure-leakage information – "Información de perdida de compresión"
Private Litigation Reform Act – "Ley de Reforma de Litigios de 1955"	Prop Strike Inspection – "Inspección de (para) impactos en hélice(s)"
Propeller feathering – "Embanderamiento de la hélice"	Propeller tie-down boots – "Cubiertas sujeción / anclaje de las palas de la hélice"
Protective masking / masking protector – "Envoltorio protector o enmascaramiento protector"	Protruding head type Rivet – "Remache cabeza alomada"

PSF (Pound per square foot) – "Libra por pie cuadrado"	Power Turbine (PT) – "Turbina de potencia"
Pull any G's – "Sufrir la inercia"	Pulse landing light system – "Sistema de luces de aterrizaje por pulsos"
Pulse/Echo type – "Tipo pulso-eco"	Pushrods – "Verillas de empuje"
QAD IDG (Quality Assurance Department Integrated Drive Generator system) – "Rápido para ata/desatar ; generador por una transmisión integrada"	Quadcopter – "Cuadricoptero, cuadrirrotor, quadrotor, drone cuadricóptero"
Quatity take offs – "Cantidad de despegues"	Racking practice – "Manejo de los estantes / racks"
Radio Frequency Guidelines – "Normas de emisión de emisión de las radiofrecuencias"	Ram – "Dispositivo de aire forzado"
Ram air turbine – "Turbina de aire de impacto"	Ramp 1 check – "Verificacion de la rampa 1"
Ramp presence – "Imponencia (en plataforma/rampa)"	Ramp Tower – "Torre de control de plataforma/rampa"
Range – "Base de lanzamiento / autonomía"	Range rings and angle marks – "Anillos de alcance y marcas de ángulo"
Rattletrap airplane – "Avión destartalado"	Raw video / raw amplitude video – "Video tal como se recibe (original) / video de amplitud tal como se recibe (original)"
Read out range and azimuth (to and from) – "Rango de lectura y azimut (hacia y desde)"	Readiness Log – "Cuaderno Bitácora de Mantenimiento"

Real beam ground mapping – "Mapeo de terreno con haz en vivo"	Real-time mission intelligence – "Información para misiones en tiempo real"
Recency – "Tiempo máximo desde promoción"	Recess – "Rebaje"
Reciprocating powered airplanes – "Aeronaves de motores recíprocos"	Recover to PO – "Recuperar (los ingresos) hasta (los términos/montos de) la orden de compra"
Red-eye flights – "Vuelos nocturnos"	Reefing line cutters – "Seccionadores de líneas de rizado"
Reflective gable marker – "Baliza rectangular (o plana) reflectante"	Remain Overnight Clean – "Proceso de Limpieza Nocturno / Limpieza por Pernocta (de Aviones)"
Remote sensing satellite – "Satélite de teledetección / satélite con sensor remoto / satélite con detección remota"	Rent holiday – "Periodo de gracia de pago de alquiler / periodo de carencia en el alquiler"
Reo – "(Estructure para) Refuerzo"	Repack cycle – "Ciclo de reempacado"
Report on watch with – "Comunicarle a que se encuentra de turno"	Reside in a vacuum – "Esta montado en un vacio parcial"
Resin Infusion – "Infusión de resinas"	Return ship via – "Sistema de devolución"
Revenue service – "Servicios generadores de ingresos"	Reverse engineering – "Descompilación"
Revisit – "Tiempo de revista (cuando el satélite pasa por el mismo punto)"	Rib station – "Ubicaciones de las costillas (en los largueros de ala)"
Right cross wind conditions – "Condiciones de viento cruzado por la derecha"	Rocket-powered aircraft – "Avión propulsado por cohetes"

Roistered – "En lista (de reserva)"	Roll and pitch – "Rollo y cabeceo"
Rolled to a sixty-degree left bank – "Alcanzo un alabeo de 60 grados a la izquierda"	Rolle wings-level – "(se) Alabeo para niverlarse"
Rotating condition free – "Estado de rotación libre"	Routing fees and overflight fees – "Cuota por utilización de ruta y cuota por utilización de rute fuera del limite"
Routing, label – "Asignación de rutas, asignador de rutas con interruptor"	RTF (Radio Transmission Frequency) cross-coupling – "Acoplamiento cruzado de frecuencia de transmisión de radio"
Rudder blocking device – "Dispositivo de bloqueo del timón de dirección"	Run-up área – "Área de pruebas"
Runway – "Pista de aterrizaje o despegue"	Runway charges – "Derechos de pista"
Runway excursión – "Se le acabo la pista"	Runway excursion – "Se le acabo la pista"
Runway heading – "Rumbo de (la) pista"	RX days – "Días de licencia médica, días de prescripción medica"
Safety bulkhead – "Mamparo de sequridad"	Safety chain pole – "Poste de la cadena seguridad o de la valla de seguridad"
Safety sock – "Protector de seguridad"	Scab patches – "Moldeado / Parche de reparación / parches de metal"
Scrapped part – "Pieza desechada / pieza condenada"	SE (Single Engine) and ME (Multiple Engine) pilot – "Piloto de aviones monomotor y multimotor"
Seal leakage drain port – "Puerto de drenado para la fuga en el sello"	Seat inventory – "Disponibilidad de asientos o disponibilidad de plazas"

Seat only market – "Mercado de venta de pasajes exclusivamente"	Seating pot – "Cajetín de luz empotrable"
Secondary outside world communication building – "Edificio alterno de comunicaciones (externas)"	Secondary rocket burns – "Ignición(es) del cohete secundario"
Seize-closed conditions – "Condiciones de trabado-cerrado"	Selectable histories – "Historiales seleccionables"
Serialised on-condition/condition monitored components inventory – "Inventario de componente seriados en condición/de condición supervisada"	Serve a hot destination – "Prestan servicio a destinos cálidos y elevados/de gran altitud"
Sextillions – "Miles de trillones"	Shear nut – "Tuerca de esfuerzo constante"
Shop cards – "Tarjetas de mantenimiento"	Shop revisión – "Modificacion de taller / revisión de taller"
Shoulder – "Envergadura"	Simulataneous offest instrument approach (SOIA) – "Aproximación instrumental simultanea offset"
Single rocket – "Un único cohete"	Single-airline websites – "Sitios web particulares de las empresas"
Sink rate – "Tasa de descenso"	Skiddometer – "Medidor de acción de frenado"
Skycap – "Auxiliar de equipajes"	Sled test – "Prueba de deslizamiento (de los asiento expulsables)"
Slide – "Rampa de evacuación"	Slip allowance – "Asignación por descanso"

Slipper seal – "Sello de patín"	Slot controlled airports – "Aeropuertos controlados por slots"
Soar to success with aerial information today – "Despega hacia el éxito con información aeronáutica mas actual"	Socket basket – "Canasta / cesta hueca"
Solar array – "Módulos / paneles solares"	Solid universal head rivet / clamp – "Remache macizo de cabeza universal abrazadera / mordaza"
Solid / liquid launch vehicle – "Vehículo de lanzamiento de combustible solido/liquido"	Sonic boom – "Explosión, boom o estallido sónico (o)"
Sonic compression – "Compresión sonora"	Spacing equipment – "Equipo de control de anchura de banda"
Sparing level – "Nivel de reposición"	Spin-up – "Centrífuga"
Split line – "La unión/ la línea divisoria"	Split line screws – "Los tornillos de la unión"
Spooldown – "Durante el apagado"	Safe Performance Self Assessment (SPSA) – "Autoevaluación del auto desempeño"
SRB (Solid Rocket Booster) Hold Down Posts – "Postes de sujeción de cohetes sólidos"	Stack drilling – "Taladrado de piezas apiladas, taladrado de chapas apiladas"
Stage separation – "Separación de estapas"	Staggered approach – "Aproximación no simultanea"
Staging units – "Plataformas para escalas aéreas"	Standard factory prototype test results – "Resultados de prototipos de ensayo de fabrica estándar"

Starlifter – "Elevador estelar Lockheed C-141"	Stealth kits – "Kit/actualizaciones antirradar"
Sterile flight deck – "Cubierta de vuelo aséptica / estirilizada"	Stick fueling – "Varilla medidora del nivel de combustible"
Stop-off Paint – "Pintura aislante / pintura anticorrosiva"	Strake patents – "Patentes de toma de aire"
Strike the rivet a short blast – "(dispón, prepara, establece la remachadora para que) de golpes cortos a los remaches"	Submittals (Engine Data submittals) – "Envíos (de datos de motor)"
SUP (Supply) Contracts – "Contratos de soport""	Sufficiently different carriers – "Portadoras suficientemente separadas"
Synchophasor – "Fasor sincronizado / Sincrofasor"	Supplemental Type Certificate – "Certificado suplementario / complementario de tipo"
Synthetic target plots and tracks – "Representaciones graficas y rutas sintéticas de los blancos"	Synthetic picture from primary radar extractor and IFF (Identification Friend or Foe) extractor – "Imagen sintética del extractor primario del radar y del extractor IFF (identificación amigo o enemigo)"
Tactical decisión aids (TDAs) – "Sistemas de ayuda a la decisión táctica (TDA)"	Syzygy – "Sizigia (eclipse solar total)"
Tail rotor drive shaft – "Eje del rotor de la cola"	Tag – "Tira / franja / fajas / fichas / tarjetas de progreso de vuelo"
Take-off thrust / reverse thrust – "Empuje de despegue / empuje de reversa"	Tail trim track – "Tanque de la cola"
Tap testing – "Prueba de percusión"	Takeoff roll – "Rodado para despegue"

Technology punch – "Avance tecnológico (en contexto de aeronáutica)"	Taxi – "Carreteo"
Telescoping escape pole – "Barra telescópica de escape"	Teflon lined nuts / castle nuts – "Tuercas revestidas de teflón/tuercas acastilladas"
Test cell run – "Prueba en la celda de pruebas"	Terminating action basis – "De manera definitiva, acción de carácter definitivo"
The canopy braking knife hook safety – "Gancho de seguridad de la navaja del tope del toldo"	Testing cart – "Banco neumático de pruebas"
The power robbing sideways lift – "La fuerza de eleva ion y la turbulencia laterales que reducen la potencia"	The ground gripper – "La tenaza a tierra"
The SSR (Secondary Surveillance Radar) was not moved – "El radar secundario no se ha movido / desplazado"	The referencing with gravity more robust – "La ubicación bajo gravedad mas confiable"
These services include the booking engines as well as back-office recordkeeping – "Estos servicios incluyen los motores de reservaciones así como el mantenimiento de archivos de la oficina regional"	The various earth-bound panels and chambers used in designing space enclosures – "Los diversos paneles y cámaras fijos usados en el diseño de recintos"
Threads of the bearing – "Roscas del cojinete"	Threaded rams – "Émbolos roscados con tuercas reten"
Thrust rating – "Potencia de empuje"	Through flights – "Vuelos directos"
Time Compliance Technical Orders (TCTOs) – Ordenes técnicas con tiempo de	Time-lapsed data – "Datos a intervalos de tiempo regulares (prefijados, etc.)"

cumplimiento"	
Timing disc – "Disco descodificador / graduador / regulador / de sincronización /temporizador"	Tip rib – "Costilla de la punta"
To be upset – "Desestabilizado"	To fail inspection – "No pasar la inspección"
To flow into orbit – "Orbitar, entrar en orbita"	To hand prop a plane – "Arrancar a mano un avión"
To over-read – "Indique/lea de mas"	To pre-clear – "Pasar, desaduanizar"
To save a sore finger from twisting drills by hand – "Evitar lastimarse los dedos por girar la broca con la mano"	To screen catering suppliers – "Efectuar el control de los proveedores de comida"
To trickle charge a battery – "Cargar continua y lentamente una batería / una carga continua y lenta de una batería"	Toe mirror – "Espejo retrovisor"
Tooling gauge – "Calibrador de herramientas (universal)"	Touch and go – "Toma y despeque"
Tow head – "Cabeza de remolque"	Tower coordinator – "Coordinador de torre"
Tracks – "Trazas"	Trade studies – "Análisis de alternativas"
Trailing edge / leading edge panels – "Paneles de borde de salida/borde de ataque"	Trailing edge flap – "Aleta/flap del borde de fuga/salida depende del destino)"

Transmit pulse is composed of 4 sub-pulses – "El pulso de transmisión se compone de 4 sub-pulsos"	Transponder Enhanced with Diversity – "Transpondedor con diversidad mejorada"
Trim dial – "Dial (mando o botón) de ajuste"	Trim tabs cabling – "Cableado de aletas equilibradoras (compensadoras)"
Trimmed out – "Compensarse / mantener la posición de equilibrio"	Trimmer – "Aleta compensadora"
Trunk flights – "Vuelos/rutas troncales"	Truss – "Armazón"
Truth in Leasing – "Veracidad en arrendamiento"	Tube fillers – "Rellenos de tubos"
Tumbleweed shimmer fabric – "Tela brillante granulada"	Tumbling dice – "Peonza tumbada"
Turnaround – "Tiempo de escala"	Twisted cable – "Cable trenzado"
Type Certificate Data Sheet (TCDS) – "Hoja de datos certificado tipo"	Type rated – "Entrenado para colar un modelo especifico de avion"
Type rating practical test – "Chequeo practico de calificación en el tipo de aeronave (la que sea del caso)"	Uncoupling the nozzle – "Desacoplando / soltando el inyector"
Under the big sky – "Bajo el ancho cielo azul"	Underslung – "Colgante"
United States principal party of interest – "Las Partes Principales Interesadas en los EE.UU."	University Aerospace Challenge Award – "Premio universitario al (mejor) desafío / reto aerospacial"

Unroll the lanyard cord and secure it to the raft or to a FA – "Cordón de seguridad"	UOM (Unit of Measure) : EA (Each) – "Unidad de medida: pieza"
Up-line station – "Estación de la que procede el vuelo"	Up-linked graphical weather – "meteorología grafica con enlace ascendente"
Uplift ratio – "Proporción/ porcentaje de alza"	Upper torso restraints – "Cinturones de seguridad de pecho/arnés de pecho"
Upscreening – "Verificaciones adicionales (para aumentar calidad o propiedades)"	Use of read back procedures – "Uso de procedimiento de colación/colacionamiento"
Valsalva maneuver – "Maniobra de Valsava"	VCCS (Voice Control Communication System) – "Sistema de comunicación por voz"
Vectoring propeller – "Hélices orientables"	VIP (Versatile information processor) configuration – "Configuración del VIP (Procesador información versátil)
Vortilons – "Vortilon (dispositivo generador de vórtices)"	Walk-on flights – "Vuelo sin reserva (de plazas)"
Walk-through – "Sin interrupciones, sin tener que detenerse"	Wash bag or toilet bag or kit bag – "Bolsa de aseo o neceser"
Wash ring – "Anillo de lavado"	Watch ítems – "Puntos/detalles (señalados/marcados) para vigilancia/seguimiento"
Water-jet thru-transmission equipment – "Equipo de transmisión por chorro de aqua"	Way points – "Waypoints"
Wear coupon – "Ficha de desgaste"	Weigh paperwork – "Documentación de peso (masa) y balance"
Weight and balance envelope – "Envolvente de peso y equilibrado"	Wet spline starter generator – "Arrancador-generador de acoplamiento por árbol estriado"

Wet-downs – "Mojar el suelo"	Wheel bay – "Compartimiento del tren de aterrizaje"
Wheel carriage – "Carruaje porta ruedas"	Wheel jack – "Elevador / gato"
Wheels off, wheels on – "Horas netas de vuelo (desde el despegue hasta el aterrizaje)"	Which is providing landing gear structure – "Que proporciona/suministra la estructura del tren de aterrizaje"
Whiskey compass – "Brújula aeronáutica"	Wide swath – "Amplio abatimiento"
Will be missile-based (launching satellites) – "Se basara en un misil (lanzamiento de satélites)"	Williams TAP (Total Assurance Plus) Elite engine programme – "Programa de motores Williams Total Assurance Plus (TAP) Elite"
Wind turbine farm – "Parque eólico"	Windmilling time – "Periodo de rotación de la hélice en molinete"
Window exit – "Ventanilla de salida de emergencia / ventanilla de emergencia"	Windowshield – "Parabrisas / ventana / luna"
Windshear and microburst events – "Eventos de cortantes de viento y eventos de microrráfagas"	Wing aspect ratio – "Alargamiento alar"
Wing carry-through spars – "Estructura del larguero del ala principal"	Wing flap – "Flap"
Wing glove – "Guante del ala o guante alar"	Wing route – "Raíz de ala"
Wing the free end of the wire – "Enrolle/doble el extremo/punta libre del cable alrededor del punto de anclaje/unión	Wing trailing edge control surfaces – "Superficies de control del borde de salida del ala"

con la herramien"	
Wing-walker – "Puntero de ala"	Wing/central wing box joint – "Unión ala/cajón (central) del ala"
Wingtip marker procedures – "Procedimientos de demarcación de extremo de ala"	Wire strands – "Filamentos de alambre"
With or without a Head Stock – Tail Stock system for revolution parts – "Con o sin sistema de cabezal/contrapunto para operaciones de torneado"	Work packets – "Paquetes de trabajo"
Xiphoid process – "Proceso xifoide"	Yaw damper – "Amortiguador de guiñada"
Yaw/roll coupling – "Acopamiento de derivación e inclinación"	Yellow "Dayglo" patches – "Parche o insignia auto reflectante amarilla"
Zero clearance or light tap fit – "Cero/sin intersticios o ajuste hermético/perfecto"	Zonal "C" check – "Revisión C por áreas"

Top Shipbuilding and Aerospace Entities / Las Principales Entidades de Construcción Naval y Aerospacial

China Shipbuilding Group – China's naval fleet/Flota naval de china.	**Boeing** – Commercial and defense air / Aeronaves espaciales comerciales y de defensa.
Daewoo Shipbuilding & Marine Engineering – Largest dock and percentage of ships/Muelle más grande y porcentaje de barcos.	**Airbus** – Commercial, defense, helicopters / Comercial, defensa, helicópteros.
Damen Shipyards Group – Warships, freighters, passenger ships, and yachts/Buques de guerra, cargueros, barcos de pasajeros y yates.	**LockheedMartin** – Largest defense and aeronautics contractor / El mayor contratista de defensa y aeronáutica.
Fincantieri – Primarily occupied with military and commercial ships of high quality/Ocupado principalmente con barcos militares y comerciales de alta calidad.	**General Electric (GE) Aviation** – Jet engine manufacturing and avionics / Fabricación de motores a reacción y aviónica.
Hyundai Heavy Industries – Largest ship manufacturing facility/La mayor instalación de fabricación de barcos	**Northrop Grumman** – Defense and aeronautical contractor for NASA, DHS, and NSF / Contratista aeronáutico y de defensa para NASA, DHS y NSF.
Imabari Shipbuilding – Evergreen G-class container ships/Buques portacontenedores Evergreen clase G.	**Raytheon Technologies Corporation** - Develops and manufactures advanced technology products in the aerospace and defence industry / Desarrolla y fabrica productos de tecnología avanzada en la industria aeroespacial y de

	defensa.
K Shipbuilding – Low temperature carriers/Portadores de baja temperatura	**United Technologies Corporation** – Aircraft engines, security, and aerospace systems / Motores de aviones, seguridad y sistemas aeroespaciales.
Mitsubishi Heavy Industries – Plans and builds larger marine vessels and platforms/ Planifica y construye embarcaciones y plataformas marinas más grandes.	**Safran** - Researches and manufactures aircraft/rocket engines / Investiga y fabrica motores de aviones/cohetes.
Sumitomo Heavy Industries – Supertankers and freighters/ Superpetroleros y cargueros.	**Rolls Royce** – Aircraft engines, missiles, and naval systems / Motores de aviones, misiles y sistemas navales.
Samsung Heavy Industries – Holds largest shipbuilding contract/Tiene el mayor contrato de construcción naval.	**Leonardo** – Unmanned aerial vehicles and helicopters / Vehículos aéreos no tripulados y helicópteros.
JSC United Shipbuilding Corporation – Russia's state-owned commercial passenger and goods vessels producer / Productor estatal ruso de buques comerciales de pasajeros y mercancías.	**NASA** – Agency responsible for Artemis, SLS, Perseverance, and James Webb Space Telescope / Agencia responsable de Artemis, SLS, Perseverance y el telescopio espacial James Webb.
Mazagon Dock Limited – Mumbai, India producer of warships, submarines, drilling platforms, tankers, cargo carriers, and	**SpaceX** – Falcon 9, Starlink, Mars colonization / Falcon 9, Starlink, colonización de Marte.

passenger vessels / Mumbai, India, productor de buques de guerra, submarinos, plataformas de perforación, buques cisterna, buques de carga y buques de pasajeros.	
Cochin Shipyard Limited – Offshore vessels, oil tankers, aircraft carriers, and repair facilities / Buques costa afuera, petroleros, portaaviones e instalaciones de reparación.	**General Dynamics** – Fighting Falcon F-16, missiles, submarines, rockets, and military services / Luchando contra Falcon F-16, misiles, submarinos, cohetes y servicios militares.
Harland and Wolff – Maker of the RMS Titanic and other cruise liners / Fabricante del RMS Titanic y otros cruceros.	**Blue Origin** – Reusable spacecraft / Nave espacial reutilizable.
BAE Systems Maritime – Military naval vessels, weaponry, and advanced military technology / Buques militares, armamento y tecnología militar avanzada.	**Virgin Galactic** – Commercial spacecraft, space tourism, and low-cost space travel / Naves espaciales comerciales, turismo espacial y viajes espaciales de bajo coste.